Secrets of the Ice

Secrets of the Ice

ANTARCTICA'S CLUES TO CLIMATE, THE UNIVERSE, AND THE LIMITS OF LIFE

VERONIKA MEDUNA

Yale UNIVERSITY PRESS
NEW HAVEN AND LONDON

for Andy and Lukas,
who are yet to see the ice

Contents

Introduction **A Land of Ice**

A century after Captain Robert Falcon Scott raced Roald Amundsen to the South Pole, another drama was unfolding in Antarctica. In early February 2012, scientists at the Russian Vostok Station were working non-stop in a race against time to complete their task before the long darkness and extreme cold of winter encroached. They had only a few days left before the last flight of the season would evacuate everybody from their desolate and remote base in the middle of the Polar Plateau. Then all radio communication ceased. After a weekend of speculation, the news arrived that the team had accomplished its mission and opened a window on one of the last untouched worlds on the planet. Lake Vostok, a giant body of water sealed deep underneath Antarctica's massive ice sheet, had finally been pierced.

This was the result of two decades of research, often amid controversy, and a feat of engineering. But rather than a conclusion, the drilling of Lake Vostok marks the beginning of a new field of Antarctic research: the search for life in subglacial lakes.

The Vostok Station was established in 1957 and soon became synonymous with hardship. The building stands on ice that runs almost 4 kilometres (2.5 miles) deep, and in all directions the view stretches across nothing but more ice. The coldest temperature on Earth, minus 89.2 degrees Celsius (-128.6 degrees Fahrenheit), was recorded at Vostok. Nobody believed back then that water could remain liquid in this deadly, frozen landscape until, in the 1970s, a British team used airborne ice-penetrating radar to survey the mountainous landscape hidden beneath this giant blanket of ice. It returned unusual readings, suggesting a massive lens of water wedged between the ice and the underlying bedrock. Later, satellite images showed a long and narrow basin, but it wasn't until 1996 that a paper published in the journal *Nature* confirmed the presence of a giant and deep lake, the size of Lake Ontario or Lake Baikal, right under Vostok Station.

Some years earlier, Russian glaciologists had already started drilling through the ice as part of a climate research project. They penetrated 3.6 kilometres

OPPOSITE During an Antarctic summer, the sun circles above the horizon but never sets.
PETER MARRIOTT, NIWA

'Ivan' the terra-bus waits to ferry
American scientists and staff from
the ice runway to McMurdo Station

(2.2 miles) deep and recovered an ice core that represents almost half a million years of Antarctic climate history. But the confirmation of a huge subglacial lake underneath their feet changed their focus. Eager to explore this primordial lake, cut off from the world above the continent's ice cap, the team continued drilling in search of life.

Pitch dark and isolated for more than 15 million years, Lake Vostok has been described as a giant laboratory for evolution. Geologists believe it may have been formed as part of an ancient rift system some 30 million years ago, when Antarctica was a much warmer place. Then, the lake was disconnected from the rest of the world when ice began to spread down from the mountain ranges. Not much would have survived this cataclysmic change, except bacteria. These masters of adaptation and survival may well still roam in this geological time capsule after millions of years of evolution in complete isolation and under extreme conditions.

The drilling project sparked concerns about the risk of contaminating such a pristine environment and for a few years the drill bit stood still while scientists focused on developing sterile technologies. In the end, the method used to avoid contamination was to let the lake water rise up the drill hole and freeze. Any sampling of this frozen plug of an ancient world to confirm whether or not it harbours microbial life will have to wait for future summer seasons, but then the Russians won't be alone any more. Lake Vostok is the largest of about 200 subglacial lakes that have been discovered under the Antarctic ice sheets and American and British scientists have launched other projects to explore these under-ice waterways. Their hope is to unlock some of the secrets of how life evolved on Earth and what it might look like on other planets.

As the Vostok drill team was probing the distant world of a lake thousands of yards below in search of microscopic life, another international team was about to change our understanding of the ecology of one of Antarctica's largest residents,

Everyone who visits Antarctica has to take part in a brief survival course. Participants learn how to walk on crampons, cross crevasses safely and arrest a fall with an ice axe. Then they must build a shelter and spend a night in it – but with spectacular views of Mount Erebus like this from both McMurdo Station and New Zealand's Scott Base few actually sleep for long. The yellow polar tents here are Antarctic-style portaloos.

the iconic emperor penguin. Using a new technique to improve the resolution of satellite images, the scientists counted the number of birds at each colony around the entire coastline of the frozen continent – and found twice as many emperors than previously thought. The results are timely and provide an important benchmark as emperors are being used as a model species to monitor the impact of environmental change in Antarctica. It may seem that every corner of our planet has been explored, but Antarctica remains a frontier of discovery and new insights into the workings of the rest of our planet.

The superlatives that have been used to describe Antarctica are familiar even to those who have never set foot on the ice. It is the coldest, windiest, driest and highest continent on Earth and the largest of the world's deserts. It is also the last place to be discovered and to submit to the presence of people.

Antarctica is the only continent without permanent human habitation, yet it may hold the key to our survival. Its ice cap holds three quarters of the planet's fresh water, its layers of ice and sediment record past climate conditions going back millions of years, and the oceans around it drive the global food chain and a giant conveyor belt of currents that transports heat around the globe. The creatures that call Antarctica home have evolved to survive in conditions hostile to life, and now the continent's hidden lakes and river systems may reveal life forms that stretch our limits of understanding how life began on Earth.

This icy wonderland is the world's fifth-largest continent and represents a tenth of our planet's landmass. It is a place of contradictions: active volcanoes erupt from a frozen landscape, ice stretches as far as the eye can see – yet it hardly ever snows – and the arid desert is surrounded by the ocean. Antarctica's shape has been described as a giant stingray, lying flat across the end of the world with its tail sticking out towards the southern tip of South America. This extension, the Antarctic Peninsula, is unlike the rest of Antarctica in almost every aspect. Its climate is less extreme and as a consequence it harbours more life. Scientists have grouped it with several smaller islands into the separate category of maritime Antarctica. The rest of the frozen land, continental Antarctica, is truly a land of ice. There are so many different types of Antarctic ice that the continent comes with its own vocabulary.

Antarctica has two ice sheets, massive frozen blankets that cover vast stretches of land and bury entire mountain ranges. The larger East Antarctic Ice Sheet is 4 kilometres (2.5 miles) high at its maximum and hundreds of miles from one end to the other. It holds most of Antarctica's ice. If it were to melt, the world's oceans

would rise by at least 60 metres (197 feet, 66 yards). Its smaller sibling, the West Antarctic Ice Sheet, sits deep below sea level and is therefore considered more dynamic. Its ice tongues are submerged in water and could break up rapidly as the ocean warms, with the capacity to raise global sea levels by 6 metres (19.7 feet, 6.6 yards). Together the ice sheets weigh down the mountainous land beneath so much that they deform the planet under the South Pole. They connect land to sea, sea to air, and – another superlative – form the largest yet simplest landform on Earth: the Polar Plateau.

At the fringes, the sheets run out into ice shelves, huge extensions that flow beyond the coast and float on the ocean. The largest of these drifting slabs of ice, the Ross Ice Shelf, is the size of France. Despite the awesome dimensions and featureless nature of ice sheets and ice shelves, they are made up of a complex mix of smaller ice formations, including ice caps, domes, streams, rivers and glaciers.

Then, with the start of each austral, or southern, autumn, Antarctica provides the stage for the world's biggest seasonal change when sea ice begins to form, first in sheltered bays and later around the entire continent. As the ocean freezes, this apron of ice can be up to 2000 kilometres (1240 miles) wide, doubling the size of the frozen continent.

Antarctica first appeared on my horizon after my husband and I moved to New Zealand in 1993. Suddenly, this vast whiteness was no longer just the odd shape at the bottom of world maps but our second-closest neighbour. The same celebrated seafarer who had circumnavigated New Zealand had also been the first to cross the Antarctic Circle when he headed into the unknown oceans in 1772. James Cook got as far as the pack ice – not just once, but three times. For three summers, Cook navigated the edge of the sea ice looking for a way in, but was defeated by the elements each time. He wrote in his journal that he would not envy the honour of whoever eventually ventured further and discovered land, and predicted that the world would not benefit from the discovery. However, he also mentioned that seals were plentiful down south, and as soon as his journals appeared in print, a flotilla of small ships headed towards the ice. Each of the sealing and later whaling ships returned with more information and new outlines for maps.

James Clark Ross, who sailed south between 1839 and 1843 commanding two warships, HMS *Erebus* and HMS *Terror*, charted much of the coastline of the continent, including the tall, flat-topped and impenetrable Barrier, now known as the Ross Ice Shelf. His mission had a strong science focus. Ross himself was devoted to marine zoology and had recruited many scientists, including Joseph

OPPOSITE The Barne Glacier descends from the western slopes of Mount Erebus and terminates in a steep ice cliff on the west side of Ross Island. It was discovered during Scott's 1901–04 Discovery expedition.
PETER MARRIOTT, NIWA

Dalton Hooker, who was the first to recognise the yellow and green stains on the underside of floes as smudges of diatom algae. Ross's crew also drew up a large collection of marine creatures from the oceans below.

The explorations of the so-called Heroic Era – a period of roughly 25 years from the end of the nineteenth century to the early 1920s – came alive for me one summer afternoon in Lyttelton Harbour in the South Island of New Zealand. I had come to interview scientists in charge of logistics for the Cape Roberts drilling project, an ambitious effort to figure out when Antarctica had turned into the white continent we know today. As I watched the ship being loaded with pale blue containers it was easy to imagine the daring departures from this harbour a century earlier. Most if not all Heroic Era voyages had a strong science element, but their focus was on exploration. Antarctica was the last frontier, seen as separate and isolated from the rest of the world.

This changed in 1957 with the International Geophysical Year (IGY), a twelve-nation scientific endeavour during which both poles were recognised for their dual role in influencing and reacting to changes elsewhere in the world. Antarctica was beginning to emerge as the planet's early-warning station and the last great wilderness.

As a result of the scientific and political success of the IGY, the participating governments created the Antarctic Treaty, which protects all land below 60 degrees South from any use other than peaceful scientific exploration. The treaty came into force in 1961, with the twelve IGY nations as the original signatories. Since then, many more nations have joined the treaty and built research stations on islands and along the fringes of the continent. Hundreds of scientists have travelled to the bottom of the world to investigate the climate, the universe, and the limits and origins of life itself. The international research effort is ongoing and culminated in the International Polar Year in 2007/08, highlighting how relevant Antarctica is to the understanding of the entire planet. The icy continent is no longer seen as a geographical oddity but a crucial part of a global climate and environment – a sensitive barometer of change that allows scientists to ask fundamental questions about life and the world around us.

This book explores many of these questions from a New Zealand perspective and with a focus on the Ross Sea region. Antarctica fuels the imagination of many people, but in New Zealand the connection is so strong that the continent is often referred to simply as 'The Ice'. Given New Zealand's geographical proximity and the fact that many historic expeditions departed from its harbours, the frozen continent is part of its national identity. A more sombre link was forged as a legacy of the 1979 Mount Erebus plane crash which killed all 257 passengers and crew.

Heroic Era expeditions were patriotic endevours, but today, Antarctica is a place of international agreements and collaborative science, ranging across all disciplines from astronomy to zoology. Several such multi-national projects feature in the following pages, including the first chapter, which explores Antarctica's climate history from its beginnings as a lush and warm chunk of a larger landmass to the white frozen vastness we know today. While geologists search ocean sediments for clues about past climates, glaciologists are analysing bubbles of ancient air trapped in the ice to trace temperatures and concentrations of greenhouse gases in the past.

Chapter Two goes in search of marine life in Antarctica. From the annual mass migration and winter breeding of emperor penguins to the chemistry of antifreeze proteins found in white-blooded fish, different survival strategies that have evolved in the extremes of Antarctica's ocean environment are examined and explained. Chapter Three moves on land to look at how Antarctica's true survivors – a few hardy animals and plants that stay put throughout the long polar night – survive freeze-thaw cycles, total desiccation and months of complete darkness unharmed.

Then Chapter Four follows microbiologists as they search for microscopic life in Antarctica's McMurdo Dry Valleys and the permanently ice-covered lakes within this spectacular cold desert. Microbes are masters of survival against the odds. While some thrive in the relatively nutrient-rich soils underneath frozen seal

OVERLEAF Antarctic life: millions of Adélie penguins feed in the Ross Sea region and come to breed in colonies on Ross Island and along the coast of Victoria Land; a white sea anemone under the ice at Cape Evans; and rich growths of colourful crustose lichens on a rock at Cape Royds.
ANTARCTICA NZ PICTORIAL COLLECTION: RACHEL BROWN, GUS McALLISTER/K002 05/06; ROD BUDD/K081 01/02; PAUL BROADY/K053 84/85

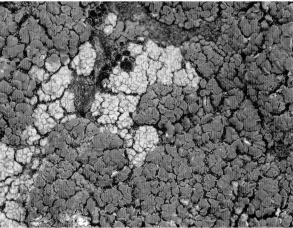

carcasses, others manage to eke out a living in sites of such depletion that their survival remains a puzzle. Now, microbes may also emerge from hidden lakes such as Vostok to reveal how life on Earth has evolved.

Finally, a brief coda delves into how Antarctica's ice is helping astronomers and physicists to capture elusive particles and to record the faint hum of the Big Bang. For all these scientific questions, Antarctica is the best – and sometimes the only – place to look for answers. Visiting this frozen landscape is to gain a fresh perspective on our world, almost like going to another planet and looking back with renewed wonder on Earth.

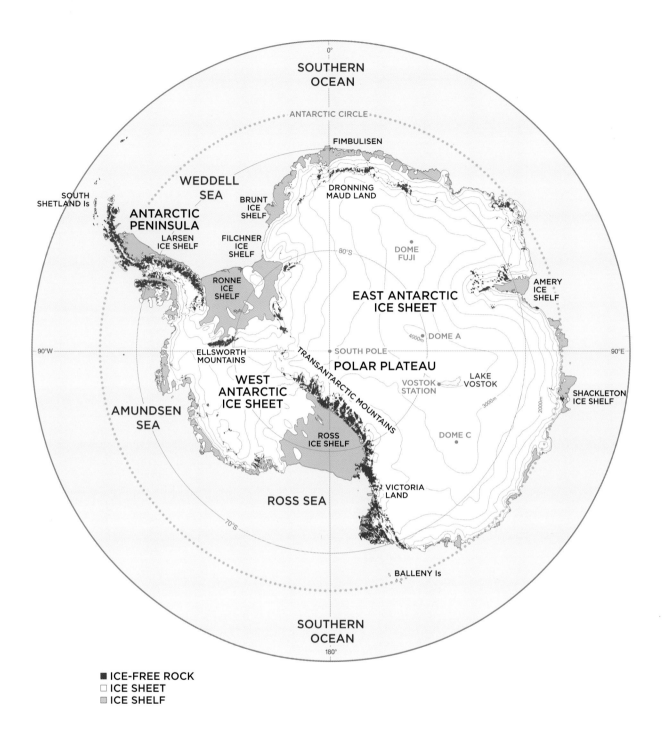

SOUTHERN OCEAN

ANTARCTIC CIRCLE

FIMBULISEN

WEDDELL SEA

SOUTH SHETLAND Is

ANTARCTIC PENINSULA

BRUNT ICE SHELF

DRONNING MAUD LAND

LARSEN ICE SHELF

FILCHNER ICE SHELF

DOME FUJI

RONNE ICE SHELF

EAST ANTARCTIC ICE SHEET

AMERY ICE SHELF

80°S

90°W

ELLSWORTH MOUNTAINS

4000m

DOME A

SOUTH POLE

POLAR PLATEAU

90°E

WEST ANTARCTIC ICE SHEET

TRANSANTARCTIC MOUNTAINS

VOSTOK STATION

LAKE VOSTOK

SHACKLETON ICE SHELF

3000m

2000m

AMUNDSEN SEA

ROSS ICE SHELF

DOME C

ROSS SEA

70°S

VICTORIA LAND

BALLENY Is

SOUTHERN OCEAN

180°

■ ICE-FREE ROCK
□ ICE SHEET
■ ICE SHELF

ABOVE The giant stingray shape of Antarctica, the land of ice.
OPPOSITE As part of the International Polar Year, GNS Science put together a new 1:250,000 geological map and accompanying GIS data set for the southern part of Victoria Land. The first geological charts of this area were drawn up by Scott's geologist Hartley Travers Ferrar during the 1901–04 Discovery expedition. Then, some 50 years later, geologists Guyon Warren and Bernie Gunn explored this region as part of New Zealand's science effort during the International Geophysical Year. The new GNS map integrates and summarises all previous geological studies in southern Victoria Land. This simplified map of southern Victoria Land is based on it. SPENCER LEVINE, AFTER SIMON COX

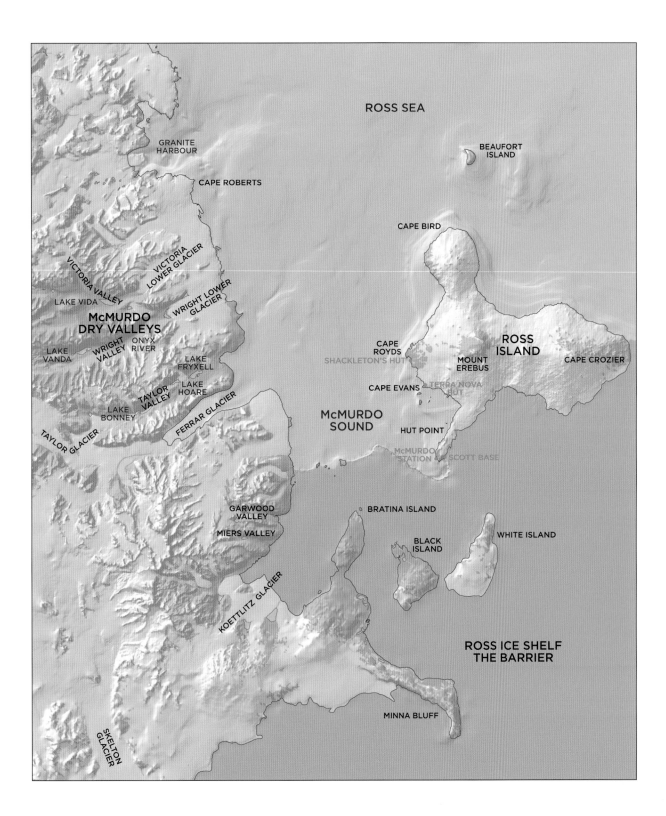

ROSS SEA

GRANITE
HARBOUR

BEAUFORT
ISLAND

CAPE ROBERTS

CAPE BIRD

VICTORIA
LOWER GLACIER

VICTORIA
VALLEY

WRIGHT LOWER
GLACIER

LAKE VIDA

ROSS
ISLAND

McMURDO
DRY VALLEYS

CAPE
ROYDS

MOUNT
EREBUS

CAPE CROZIER

LAKE
VANDA

WRIGHT
VALLEY

ONYX
RIVER

SHACKLETON'S HUT

LAKE
FRYXELL

LAKE
HOARE

CAPE EVANS

TERRA NOVA
HUT

TAYLOR
VALLEY

LAKE
BONNEY

McMURDO
SOUND

FERRAR GLACIER

HUT POINT

TAYLOR GLACIER

McMURDO
STATION & SCOTT BASE

GARWOOD
VALLEY

BRATINA ISLAND

MIERS VALLEY

BLACK
ISLAND

WHITE ISLAND

KOETTLITZ GLACIER

ROSS ICE SHELF
THE BARRIER

MINNA BLUFF

SKELTON
GLACIER

1
Uncovering the Past

Mount Erebus, here shrouded in cloud, provides a
backdrop for Scott Base. ROB McPHAIL

On the 11th of January, 1841, . . . the Antarctic Continent was first seen, . . . rising steeply from the ocean in a stupendous mountain range, peak above peak, enveloped in perpetual snow, . . . which as the sun's rays were reflected on it, exhibited a scene of such unequalled magnificence and splendour as would baffle all power of language. – James Clark Ross, *A Voyage of Discovery and Research in the Southern and Antarctic Regions*

Antarctica's most striking characteristic is its scale. The immensity of the landscape is hard to fathom, and space and distance are beyond comprehension. Viewed from Scott Base, the summit of Mount Erebus, the world's southernmost active volcano, looks like a day's walk away. Yet it is 70 kilometres (43 miles) in the distance and at 3794 metres (12,447 feet) surpasses any peak in New Zealand. The craggy horizon line of the Transantarctic Mountains 100 kilometres (60 miles) away seems within reach. So does the flat vastness of the Polar Plateau, but one only needs to read Captain Scott's diaries to appreciate the hardship involved in scaling the Ferrar Glacier to reach it.

The reason for this distorted perception is the crisp, clear air and a complete lack of familiar reference points in the landscape. A thick blanket of ice rises to more than 4 kilometres (2.5 miles) above sea level – high enough to bury all but the tallest mountain peaks and to smooth the contours of the underlying ranges. However, as permanent as it looks, this icy mantle is a relatively recent adornment. Antarctica was once wrapped in lush forests, teeming with dinosaurs and other creatures, but with the exception of a few hardy survivors, this all disappeared as the continent began to cool and freeze over. Now, this white vastness hides the most remarkable dimension of Antarctica's scale: its ancient climate history.

Underneath and within the icy surface lies one of our planet's best-kept archives of past climate conditions, stretching back millions of years beyond the earliest stirrings of human evolution or even the demise of the dinosaurs. The ice, the rocks and the ocean floor around Antarctica all tell the story of an ancient world and its many cycles of transformation from a temperate to an icy climate and back again, making Antarctica a perfect place to study the natural climate fluctuations of the past.

But this austere continent is not just a passive bystander. Its cycles of cooling and warming are also potent drivers of climate change. Antarctica's crystal

Glaciologists set up camp on the Mount Erebus saddle
to study the climate history recorded in snow and ice.

cap stores almost three quarters of the world's fresh water, frozen in a delicate equilibrium, and it powers the ocean currents that transport heat, salt, dissolved gases and nutrients around the globe. In the past, Antarctica has both reacted to and driven changes in climate, and geologists now look to the only uninhabited continent for answers when they try to predict the planet's future in the face of human-induced climate change. By drilling deep into the ice and the ocean floor around Antarctica, they recover the past in order to anticipate the future.

Gondwanan origins

Looking out towards the Polar Plateau from the short road that connects New Zealand's Scott Base and the American McMurdo Station, it is difficult to imagine that this is a living landscape and that, despite its solidity, the ice is in constant motion. This indiscernible movement and the signature it leaves behind is what scientists read, like forensic detectives reconstructing a crime from traces left at the scene. They use indirect clues from the geological and glacial record to infer what Antarctica looked like long before the first apes tried to walk upright. As snow falls on the surface, each layer is gradually compacted into solid ice and glaciologists have learned to decipher this frozen data bank to gauge conditions at the time. Similarly, but on a much longer geological timescale, the layers of sediment beneath the seafloor surrounding the continent record the grinding flow of ice sheets. Such sediment records are like ancient history books to geologists.

Scott, Shackleton, Amundsen and many other explorers of the Heroic Era recorded the first observations of Antarctica's weather conditions around the turn of the nineteenth century. After that, it took half a century for the next determined effort to begin unlocking the secrets of the polar regions during the International Geophysical Year, or IGY, in 1957/58. The twelve-nation collaborative venture focused on understanding this most inhospitable land and its links, then as yet unknown, with the rest of the planet. It also marked the beginnings of Scott Base.

The idea of a New Zealand scientific base in Antarctica was first floated in 1953, when the British explorer and geologist Vivian Fuchs announced his plans for the Commonwealth Trans-Antarctic Expedition, the first overland crossing of the southern continent. Fuchs intended to start the frigid 3473-kilometre (2158-mile) trek from the Weddell Sea and to travel via the South Pole to McMurdo Sound. He asked Edmund Hillary, freshly back from his ascent of Mount Everest, whether he would lead a group to lay food depots between the sound and the pole. After initial misgivings about the long absence required, Hillary agreed.

A year before the expedition team set sail, Trevor Hatherton, the leader of New Zealand's science team preparing to take part in the IGY, travelled to Antarctica to find a suitable spot for a base and a good route to and from the Polar Plateau. Since Scott's party had man-hauled up the Ferrar Glacier during the 1901–04 Discovery expedition, this was considered the best route. On closer inspection, however, the glacier turned out to be heavily crevassed and impassable for dogs and tractors. It fell to Hillary to survey the Skelton Glacier and to find a new piece of Antarctic real estate for Scott Base. He chose a site 2 kilometres (1.2 miles) south of the newly constructed McMurdo Station, with direct access to the Ross Ice Shelf.

The huts Hillary's men built sheltered not only their own group but also the IGY science team, and geological surveys were part of the expedition even before the Southern Tractor Party took off to lay the first food and fuel depots for Fuchs's party. Then, against instructions from the Ross Sea Committee, Hillary led a 'dash to the pole' and, on January 4, 1958, his team became the first party to reach the South Pole overland since Scott in 1912. Internationally, Hillary faced some criticism for allegedly putting adventure before the expedition's scientific aims, but in New Zealand this merely reinforced his status as a national hero.

Hatherton was passionate about science, and his enthusiasm for the international venture was such that New Zealand launched a comprehensive research programme on ice, including a study of ice thickness which produced new estimates of Antarctica's total ice content, improved meteorological predictions, and advanced the theoretical analysis of glaciers.

During those early years, Peter Barrett, who would later set up the Antarctic Research Centre at Victoria University in Wellington, was more interested in caves than in polar research. He was planning a doctoral project in speleology and was looking forward to summers spent spelunking, but an Antarctic field trip and a serendipitous discovery launched a lifelong connection with the frozen continent.

Decades later, as the quietly spoken geologist contemplates his retirement project to make his home independent of fossil fuels, he is reluctant to accept the epithet 'supremo of Antarctic geology' bestowed on him by colleagues and students. The combination of rugged adventure and scientific puzzle first drew him to the ice, he says, then Antarctica's serene beauty kept him coming back, and then he was hooked for good as he realised that this ice ball at the bottom of the world played an important part in making the rest of the planet inhabitable.

Barrett's first expedition in 1962, as an assistant with a University of Minnesota party exploring the Ellsworth Mountains in West Antarctica, took him to hitherto unexplored places. The group of eight was transported in by the first ski-equipped C-130 Hercules to operate on the continent. They lived like nomads, sleeping in Scott polar tents (pyramid-shaped shelters patterned after the tents designed by the British explorer a century ago) that had to be packed onto motor toboggans every few days to allow the team to map the area's geology over a distance of 800 kilometres (500 miles) in three months. They spent their days climbing ridges, recording and sampling. Their main task in the evenings was to try contacting McMurdo Station with a suitcase-sized heavy radio that worked mostly but not always. This first geological survey of the region found that the mountains were really a fragment of East Antarctica, torn out of place.

The following season Barrett teamed up with his University of Auckland room-mate Vic McGregor and two other young New Zealanders to survey an unexplored section of the central Transantarctic Mountains, the massive spine that runs between Antarctica's two main ice sheets. This time, their only means of transport in and out of the field was a DC-3 fixed-wing aircraft flown by American pilots who had honed their skills in the Vietnam War and could land the plane almost anywhere. There were no maps, but the aerial photographs taken in 1946 during Operation Highjump, a US Navy programme that preceded the construction of McMurdo Station, were invaluable. Barrett remembers studying the images so intensely before his first flight out into the field that he felt almost instantly at home in the icy landscape, as if he were visiting an old friend.

Once he and his team-mates, their sledge, motor toboggan, tents and food boxes had been dropped off, they were left largely to their own devices. It took two hours to make breakfast in bed each morning, with a kerosene-fuelled primus buzzing

Geologist Peter Barrett, lying on 260-million-year-old mudstone in Camp Valley in the Allan Hills, measures the orientation of ridges that are just visible on the surface of an embedded boulder. The boulder had most likely been carried by floating ice and dropped into lake mud, he found, with the furrows forming as it was dragged along beneath a Permian glacier. JANE FRANCIS

away to melt ice for a cup of tea, porridge and a bacon omelette. But they were certainly not starving. Chocolate for lunch and dehydrated minced or curried beef and potato flakes for dinner provided twice the amount of calories they would need off ice.

During this second trip, Barrett was captivated by the spectacular horizontal layers of Beacon sandstone that make up the top 2000 metres (6600 feet) of this massive mountain range, and the remarkable detail this geological formation preserves of deserts and plains hundreds of millions of years ago, at a time when Antarctica was still part of the ancient southern supercontinent Gondwana. Scott's geologists had already established that these strata span the period during which trees developed and flourished to form coal swamps. Today, however, this area is so cold and poor in nutrients that it is considered the closest analogue on Earth to Martian conditions.

Barrett's next two expeditions between 1966 and 1968 were as an Ohio State University doctoral student, travelling with the US Antarctic Program to study Beacon sandstone strata in more detail in the Beardmore Glacier area. It was midway through the second season, high up in the mountains at the head of the glacier – close to the South Pole route that both Scott and Shackleton had followed more than half a century earlier – that Barrett made his first major discovery. Crawling along an ancient riverbed to check a pebbly layer, he spotted a small

Beacon sandstones are a legacy of Antarctica's warm past; while further down the slope dark intrusions of fire-formed Ferrar dolerites are a signature of even more ancient geological processes. ROB McPHAIL

stone that looked out of place. It turned out to be a fragment of fossilised jawbone from a salamander-like creature that lived during the early Triassic period, at a time when dinosaurs dominated the world. This tiny bone helped to clinch one of the big scientific debates of the time.

The fossil fragment provided the first evidence that land vertebrates had roamed Antarctica when it was a much warmer place. What's more, subsequent discoveries by other palaeontologists included bones of a reptile species known to also occur in South Africa and China. These finds provided solid evidence in support of two ideas that were still controversial at the time: continental drift and the existence of a Gondwana supercontinent.

Tracking the deep freeze

Peter Barrett went on to lead many other scientific expeditions to Antarctica and has spent a considerable portion of his life on ice. But, even decades later, the sense of adventure and isolation of these early field seasons remains a precious memory. Over the years, he became part of an international group of eminent geologists who helped unravel Antarctica's geological and climate history, from the moment it started its journey as a separate continent to its transformation into the frozen whiteness we know today.

When Antarctica was still part of Gondwana, it was a green continent with rivers, plains, forests, amphibians and dinosaurs. The evidence is still preserved in exposed rock faces along the Transantarctic Mountains. Ripples and other patterns etched in the sandstone reveal the flow and strength of ancient watercourses. Coal beds and even tree stumps tell of a more pleasant climate. And there are the bones of reptiles and amphibians that thrived amidst these ancient conditions.

Back then, Antarctica wasn't in its present position. It drifted towards the pole slowly, over millions of years, after breaking free from Gondwana about 180 million years ago, just a geological moment after the appearance of the first shrew-like mammals. As the supercontinent was breaking up and its fragments began drifting away from each other, the ancient oceans spilled into the new gaps between landmasses and formed the South Atlantic. This profoundly changed the global pattern of ocean currents, illustrated most spectacularly by the opening up of the Southern Ocean. Once Antarctica had reached its polar position, the ocean formed a strong and deep current that was circling around the continent and isolating it even further from other land.

Antarctica's polar position and this great circumpolar current alone may have been good explanations for why the continent became so cold that ice grew all over it, but the geological record shows that the journey to the pole was completed millions of years before any major ice sheets began to spread. Scientists were faced with a dilemma and the challenge to figure out which other factors were influencing Antarctica's climate. Surveys of the few small areas of exposed rock and debris left behind by glaciers suggested that the continent's massive ice sheets were no older than a few million years, but new exploration technology was about to paint a clearer picture.

During the post-war period, the search for oil was expanding to most areas along the planet's continental shelves, and echo-sounding and seismic techniques were developed to survey the seafloor. Geologists realised that the ship-based oil-drilling technology could also be useful for studying the Earth's crust, and by the mid-1960s, American marine scientists had managed to persuade the US administration to support an international project to drill into the bottom of the ocean. The main goals of the project were to test the still new theory of plate tectonics and to study the largely unknown oceans, but Barrett and a number of his colleagues also realised that studying seafloor sediments would be the only way to find out how and when Antarctica's ice cover first appeared.

In December 1972, Barrett and two other New Zealanders, geologist Peter Webb and palaeontologist Derek Burns, climbed on board the *Glomar Challenger*, the first research vessel equipped for deep-sea drilling, to test the idea during a leg of the Deep Sea Drilling Project in the Antarctic region. Webb had already achieved a degree of notoriety in the Antarctic community by this stage. Back in 1957, he had celebrated his 21st birthday on board the HMNZS *Endeavour*, heading to Antarctica with the team for the second season of the Trans-Antarctic Expedition and the International Geophysical Year. He and his friend Barrie McKelvey, both third-year geology students at Victoria University at the time, had managed to secure a place on the understanding that they would help unload the cargo at Scott Base and return home promptly once the job was done. But they found that geologists were in demand at McMurdo Station and joined American parties on geological excursions. As Webb puts it, the pair spent the rest of the summer avoiding repatriation by heading as far into the mountains as possible – mapping and describing virgin territory in the process.

The *Glomar Challenger* voyage marked the beginning of a lifelong collaboration between Barrett and Webb, by then a geologist with the New Zealand Geological Survey. The ship sailed south from Fremantle and drilled several holes into the ocean floor along a transect, or line, between Australia and Antarctica. A drill site in the middle of the Ross Sea first unearthed recent layers of mud that were, not surprisingly, clearly glacial deposits, but Barrett recalls his disbelief when layer after layer showed the same material, right back to deposits that were 25 million years old – or some ten times older than anybody had expected the Antarctic ice sheets to be.

A few months later, the next cruise was led by Jim Kennett, a New Zealander who had moved to the US as a freshly graduated geologist to become a professor at the University of Rhode Island. Kennett knew early on that he wanted to study the oceans as a way of tracking Earth's history. In fact, he had his research career mapped out at the age of eleven. By then, he had already filled a spare room in his family home with a collection of rocks, shells, bones and fossils. He had also been given a sample of microscopic fossils called foraminifera – or forams, to those who work with them. Forams come in many species, extinct and alive, which allow geologists to date the layers of ocean sediment. Later, as a student, Kennett realised that these tiny fossils also carried a lot of information about the environmental conditions of their time and so, for his doctorate, he chose to study sediments that had been laid down four to seven million years ago all across New Zealand. Since ocean-drilling technology wasn't available then, he travelled the country on his motorcycle to study marine sections that had been thrust up on land by earthquakes. To work out the temperatures at the time, he looked for foram species he knew preferred either cold or warm conditions.

By the time he sailed south as part of the Deep Sea Drilling Project, Kennett had already spent time on ice, like Barrett and Webb, helping to map some of the last uncharted territories on Earth. This time his destination was the bottom of the deep ocean south of Australia and New Zealand where he hoped to retrieve a complete record of strata representing the past 65 million years. The mission was successful, and the core also brought up layers of mud with well-preserved forams, allowing Kennett to use isotope analysis, a technique used widely today but at the cutting edge of science in those days. He worked with Nicholas Shackleton, a relative of the British explorer Ernest Shackleton, who was using the same technique to analyse ocean sediments from other drilling projects at his laboratory at Cambridge University.

Isotopes are versions of the same element which differ only in their weight. While they are chemically the same, their physical properties are different. Some of the most common elements that make up living organisms – hydrogen, oxygen and carbon – consist of at least two isotopes, and the ratio between the two versions is like a signature of environmental conditions. Oxygen isotope ratios found in fossil forams, for example, can be used to determine the temperature of the ancient waters they once lived in. The heavier isotope doesn't evaporate as easily as its lighter cousin, and so the ratio also reflects the amount of ice locked up in ice sheets and glaciers.

With the sediment cores from the Southern Ocean, Kennett's crew extracted intricate microscopic calcium carbonate shells left behind by forams millions of years ago, exquisitely preserved in the mud. These shells contain oxygen isotopes and Kennett used them to reconstruct past climate conditions and ocean temperatures. His results showed two clear shifts in the isotope record, both of which are still valid today. There was an abrupt increase in the heavier oxygen isotope around 34 million years ago as a consequence of global cooling and extensive sea-ice growth around Antarctica, and another similar event around 14 million years ago. Scientists already suspected the earlier climate shift because the geological record showed a marked change in terrestrial vegetation around the world, but the isotopic record provided further support for the proposition that the world's oceans had cooled significantly some 34 million years ago. This point in time is now understood as the beginning of the formation of the first large Antarctic ice sheets, and the later event, 14 million years ago, heralds the development of an even larger and more stable ice sheet as we know it today.

Pacemakers of ice ages

A few years after the *Glomar Challenger* voyages, Nicholas Shackleton set out to test a hypothesis which postulated that major climate fluctuations in Earth's history had been triggered by small periodic changes in its orbit around the sun.

More than fifty years earlier, the Serbian mathematician Milutin Milankovitch had spent his internment during the First World War investigating the rhythm of these planetary wobbles. As a prisoner of war, he had been granted permission to live in Budapest and to use the libraries of the Hungarian Academy of Sciences to continue his scientific interests. Not long after the war, he published his remarkable findings. He had identified three different cycles – repeating every 100,000 years, 41,000 years and 21,000 years, and ruled by small changes in the Earth's elliptic orbit around the sun, its tilt and axis of rotation respectively – that changed the amount and distribution of solar energy the planet received. Milankovitch postulated that these cycles were acting as a metronome, setting the timing for Earth's ice ages. Shackleton began to look for physical evidence against which to test the idea.

He chose two sediment cores that had been drilled in the deep subantarctic ocean between Africa, Australia and Antarctica. Shackleton also focused on forams and the oxygen isotope ratios in their shells and, based on this painstaking analysis, showed that oscillations in climate over the past few millions of years could indeed be correlated with Milankovitch cycles, and that they were the pacemakers of the ice ages.

Combined results from several drilling projects have since helped to refine the timing of major climate shifts, but in broad terms the story Kennett and Shackleton told in the 1970s is still the same today: large, dynamic ice sheets first spread across the Antarctic continent about 34 million years ago and then, around 14 million years ago, Antarctica cooled even further and the larger of its ice sheets, known as the East Antarctic Ice Sheet, became stable.

OPPOSITE Surveying and preparing the drill rig at Cape Roberts, 1999.
ANTARCTICA NZ PICTORIAL COLLECTION: TIM NAISH/K001 99/00; K001H 05/06

However, sediments from the deep ocean only go so far in unravelling Antarctica's past. A more precise picture comes from sediment cores extracted closer to Antarctica, or the continent itself, which record the sequence of ice advances and retreats across the land and into the sea.

At about the time the Deep Sea Drilling Project was being developed, the US, New Zealand and Japan were planning a project to drill deep into the floor of the McMurdo Dry Valleys, an oasis of ice-free land in Antarctica's frozen desert. Peter Webb was the project's geological adviser. The team drilled fourteen holes during the early 1970s and unearthed a striking record. Both the geological and biological traces in the sediment reflected periodic changes in the extent of Antarctica's ice sheets, with glaciers advancing and retreating several times over the past five million years, with corresponding changes in sea levels by tens of yards. Webb teamed up with Peter Barrett to drill a fifteenth hole and continued to campaign for more, in the hope of reaching back in time to the very beginnings of Antarctica's glaciation. Their efforts culminated in the Cape Roberts Project, a seven-nation collaboration, during the late 1990s.

Unearthing the future

While geologists were sifting through the mud for the hard data, there was another, far-sighted strand of research being undertaken at Ohio State University's Institute of Polar Studies (now known as the Byrd Polar Research Center), where Peter Barrett was working on his doctoral project. Glaciologist John Mercer had joined the institute during the early 1960s after a major effort to compile an atlas of Antarctic glaciers. His special interest was in how the history of glaciers in Antarctica and South America could be deciphered from moraines on land. As a side project he pondered the past behaviour of the West Antarctic Ice Sheet, the smaller of Antarctica's two ice sheets, based on new data showing that most of it was grounded almost 200 metres (660 feet) below sea level. He also noticed that measurements at the South Pole were showing a rise in carbon dioxide levels in the atmosphere, mirroring similar increases observed in other parts of the world. Putting all of this information together, he issued one of the first warnings that humanity could wreak havoc on the world's climate if global consumption of fossil fuels continued unabated. He predicted that atmospheric concentrations of carbon dioxide would double in 50 years and result in a greenhouse-warming effect strong enough to make the West Antarctic Ice Sheet vulnerable. In 1978, he foresaw (accurately, as it was to turn out) that the floating ice shelves that buttress this ice sheet would be the first to collapse, on either side of the Antarctic Peninsula.

Barrett remembers that there was significant disagreement among glaciologists at the time about both the buttressing hypothesis and the immediacy of the threat, but Mercer's paper provided the impetus for geologists to think of greenhouse gases such as carbon dioxide as drivers of climate and to go out looking for a more detailed geological history of both the East and West Antarctic ice sheets.

With new drilling technology, Barrett as the chief scientist and Peter Webb as the US co-ordinator, the Cape Roberts Project completed three drill holes, down 1500 metres (4900 feet) into the sediment off the Antarctic coast. The team recovered a remarkable sedimentary record tracing the history of ice-sheet changes and fluctuating sea levels during a window in geological time between 17 and 34 million years ago. It confirmed that massive ice sheets covered Antarctica around 34 million years ago and that they were extensive even at this early stage, about a million years after they first appeared. But the sediment layers also showed that the ice was highly dynamic, coming and going in response to climate changes paced by Milankovitch cycles.

When Antarctica's ice cap was still young, the glacier tongues that reached the coast were regularly calving off icebergs into the Ross Sea. During this warmer period, the climate was pleasant enough for low woodland forests to thrive along the coast. But then, some 10 million years after the ice first began to form, temperatures dropped further and only low-growing, sparse tundra survived. Eventually, the ice sheets extended their chilly grip out beyond the coast and even the permafrost vegetation disappeared.

While the Cape Roberts team was able to describe changes in Antarctica's ice cover, climate and sea levels along its coast during the early years of the deep freeze, the next big question was to figure out what the continent looked like during its more recent past, particularly after the period of cooling. One obvious place to look was the Ross Ice Shelf, the largest slab of floating ice in Antarctica and the biggest and most sensitive indicator of climatic fluctuations in the region.

Glorious green mud

My initial glimpse of the seafloor landscapes underneath this giant hunk of ice came during my first Antarctic visit in 2001. Tim Naish, now the director of Victoria University's Antarctic Research Centre, was preparing for the ambitious Antarctic Geological Drilling project, or ANDRILL, whose aim is to investigate the

For its first drilling season in 2006, the ANDRILL rig was set up on the ice shelf behind Ross Island.

factors that have triggered changes in ice cover during Antarctica's geologically recent past. But before the drilling could start, one of many preparatory tasks was to find the best place to set up the drill rig by imaging seafloor sediments and ash layers several kilometres below the seabed.

Suntanned and stubbled, Naish had set up camp out on the ice. Operating from a few simple Scott polar tents, his team was drilling 2-metre (6.5-foot, 2.2-yard) holes along transects on the McMurdo Ice Shelf and laying seismic cables along the lines to connect geophones that had been buried in each hole. Geophones are small listening devices that convert ground tremors into voltage, and in this case they were used to trace the sound waves produced by the blasts detonated by more than 3 kilograms (6.6 pounds) of explosives. The echo bouncing back off the layers of rock underneath the ice shelf produced an image of the landscape below – the only way of visualising the lay of the land under this massive expanse of ice.

A few years later, I returned to Antarctica for ANDRILL's first season. Naish was now co-chief scientist, together with Northern Illinois University geologist Ross Powell. On a good day, the 40-tonne (44-ton) drill rig was clearly visible from Scott Base out on the ice shelf, and a regular shuttle service ferried personnel and freshly

Tim Naish, Stuart Henrys and Gary Wilson, three of the scientists involved with ANDRILL, during seismic site surveys. ANTARCTICA NZ PICTORIAL COLLECTION: TIM NAISH/K001 05/06

The drill rig was within sight of both Scott Base and McMurdo Station during ANDRILL's first season. ANTARCTICA NZ PICTORIAL COLLECTION: ERIK BARNES/ K400 07/08

Setting up ANDRILL's rig.
ANTARCTICA NZ PICTORIAL
COLLECTION: TIM NAISH/
K001H 05/06

unearthed cores between the base and the drill camp. In the evenings, often some hours after dinner time, the day shift of drillers and scientists turned up at base, their faces looking almost as gaunt and exhausted as those of the men from Heroic Era expeditions a century earlier. They had just handed over to the night-shift drillers, who would keep the diamond-studded drill bit going for the next twelve hours.

To get to the sediment on the ocean floor, the drillers first had to core their way through more than 80 metres (260 feet, 90 yards) of ice and then lower the drill through more than 800 metres (2600 feet, 900 yards) of super-cold ocean water – and all of this had to happen while the ice shelf was groaning and moving with the swells and tides.

Despite such logistic challenges, ANDRILL's first drill site had been selected on the ice shelf behind Ross Island because it was well placed to record glacial advances, when the ice shelf thickens to ground on the seafloor, and retreats, when it gives way to sea ice or open ocean. It was also close to the junction between East and West Antarctica, with the Transantarctic Mountains to the west splitting the continent into two unequal parts. The larger East Antarctic Ice Sheet is considered the more stable part of the ice cap, while geologists now echo Mercer's predictions and think of the West Antarctic Ice Sheet as more dynamic and mobile because most of it is grounded far below sea level.

With the help of the seismic profiles collected during earlier seasons, the team had also chosen the site because it sits in a deep-water basin that had formed around the volcanic Ross Island. The load of the rocks spewed out by Mount Erebus over the past five million years had weighed the Earth's crust down and created a moat basin deep enough to trap sediments that have faithfully recorded the advances and retreats of the giant Ross Ice Shelf.

By the time I arrived at the drill site, the team had already penetrated 700 metres (2300 feet, 765 yards) deep into the sediment, and each core I watched coming up to the surface was more than five million years old. Even to the uninitiated eye, the layers of different types of sediment were distinct – and watching the drillers handling each core, it was clear that these were very precious bits of dirt.

As soon as the cores came up to the surface, they were carefully cut into 1-metre (3.2-foot) lengths and prepared for the first measurements. Physicists traced fractures within the core to decipher the history of ancient mountain-building forces and tensions within the Antarctic tectonic plate. After these initial measurements, the core was split along its length – with one half free for sampling by several teams and the other half shipped to a drill core archive at Florida State University.

The effort to decode the information stored in the sediment layers was a major exercise in forensics. Every morning, the scientists performed the same ritual. Overnight, a team of sedimentologists, led by Ohio State University's Larry Krissek, had identified the clearest and most interesting layers in the core; and after a handover meeting in the morning, everybody lined up around a long table at McMurdo Station's Crary Lab to inspect the bounty of the last drilling shift. Quickly, each team stuck tiny flags into the core to mark out territory for sampling. Then everybody began scraping off and cutting out bits for further experiments, always keeping track of the exact location on the core so that all results could later be collated into one comprehensive record of the area's geological and climate history.

While one group was busy squeezing the last drops of water from the rocks to measure its chemistry, another team looked for any remains of fossilised shells, and yet another sifted through the layers in search of volcanic ash to help date the core by known volcanic eruptions. Traces of ancient life, which emerged throughout the core, unearthed the most compelling evidence for past changes in environmental conditions.

CLOCKWISE FROM TOP LEFT The drill team at the controls that guided the drill hundreds of yards deep into the ocean sediment. After millions of years underground, cores of mud and rocks emerge from the drill hole for inspection by scientists. The team scours the precious mud for traces of ancient life, using tiny flags to mark areas of interest. TOP RIGHT: ANTARCTICA NZ PICTORIAL COLLECTION: ERIK BARNES/K400 07/08: ALL OTHER IMAGES: PETER WEST/NSF

Whenever the core turned a tinge of green and softer than previous material, everybody's eyes lit up, but nobody was more excited than Reed Scherer, a palaeontologist from Northern Illinois University. This was what he had come for, but he had expected a few short sequences, not tens of yards of olive-coloured mud. Hunched over a microscope and calling out a string of taxonomic names to his colleagues, Scherer was brimming with almost childlike joy, certain that he was looking back at a world of open oceans and algal blooms around Antarctica.

Through the lens of his microscope, he was seeing the beautiful and intricate shells of a group of marine algae called diatoms. When these single-celled plants die, their delicate silicate shells drop to the bottom of the ocean where they are usually mixed up by tiny bottom-dwelling organisms that spend their lives digging up the seafloor. Most diatoms disappear without a trace, unless there are so many of them that they literally rain down and the seafloor becomes low in oxygen, which limits the bottom-dwellers and results in finely preserved layers of diatom remains.

With those first few hundred yards of sediment, the team had travelled back in time by about five million years through to a period known to geologists as the Pliocene, when the very first ape-like creatures began to practise standing upright. From geological evidence collected elsewhere, scientists already knew that this period was warmer, with sea levels up to 20 metres (66 feet) higher than today, average global temperatures two to three degrees above today's, and atmospheric concentrations of carbon dioxide similar to the level we are about to reach this decade. The geological window of the Pliocene is often used as an analogue for the future, and one of the goals of the ANDRILL project is to understand how Antarctica changed during this time in order to predict how the continent's ice sheets might behave during projected climatic changes over the next decades and centuries.

This was the first time anybody had seen the signature of Antarctica's moving ice sheets and ice shelves so clearly. To Scherer, the long stretches of sediment core with thick layers of compacted algal ooze – diatomite to geologists – were a clear signal that there was no Ross Ice Shelf at the time, and that much of the West Antarctic Ice Sheet had probably also retreated or even disappeared.

After months of further analysis in laboratories throughout the world, the first ANDRILL core has confirmed the scientists' early interpretations. The area that is covered by the world's largest ice shelf today fluctuated between open ocean conditions, which lasted for up to 200,000 years at a time (roughly as long as it took us to evolve from our nearest hominid ancestors), and much colder periods when the ice shelf was grounded, extending the ice sheet out far beyond the present coastline.

Over the past five million years such transitions happened not once but at least 40 times.

A year after the first drilling season, the team drilled a second hole at the southern end of McMurdo Sound. This site had been selected to reach further back in time to a period known as the middle Miocene, which provides another ancient window on a warmer world. This period is thought to have played a crucial part in the development of the Antarctic ice sheets as we know them today. It covers a change from a warmer climate about 17 million years ago to the onset of the major cooling 14 million years ago that consolidated Antarctica's growing ice cap to produce the frozen outline delineated on modern maps of the continent. Fossils and sediments deposited in this second drill core confirmed warmer-than-present conditions over an extended period of the Miocene, when the Ross Sea looked more like southernmost South America. However, among the 1107 metres (3632 feet, 1210 yards) of sediment core, a 2-metre- (6.6-foot-, 2.2-yard-) thick layer stood out. It featured an unexpected abundance of marine and terrestrial microfossils, including a fivefold increase in freshwater algae and almost 80 times as much plant pollen – suggesting a short period when Antarctica became much warmer, abruptly, just before it entered the long deep freeze about 14 million years ago.

The next phase of the project, planned for Coulman High about 120 kilometres (75 miles) from McMurdo Station on the Ross Ice Shelf, will focus on time zones that stretch back even further, overlapping and extending beyond the geological period described by the Cape Roberts Project, to a time when carbon dioxide levels were three to five times higher than today.

Any variation in the mass of Antarctica's ice sheets directly affects sea levels, but the sheer size of the continent presents a considerable challenge for scientists trying to establish how much ice is currently melting. Satellites have been used since 1979 to track Antarctica's size as it doubles every winter when sea ice begins to grow around it, but more recently, they have also been employed to survey the entire continent and to monitor any changes in mass of its icy mantle. In 2002, the Gravity Recovery and Climate Experiment (GRACE) satellites began drawing a comprehensive gravity map of the frozen continent.

One complicating factor is that the earth buckles under the weight of Antarctica's ice sheets. Whenever some of the ice melts, parts of the bedrock bounce back – and Ohio State University earth scientist Terry Wilson is using this rebound effect to make the estimates of ice loss more accurate. She is part of a project called POLENET, a vast network of Global Positioning System (GPS) and seismic sensors, which record the movement of ice and the bedrock beneath it and help to fine-tune the calculations.

At the University of Texas, geophysicist Donald Blankenship has been operating an airborne radar system to provide a bird's eye perspective on the flow of ice across Antarctica. Initially, his focus was on the smaller West Antarctic Ice Sheet because most of its ice is grounded below sea level, but since 2008, his team has been flying a small aircraft fitted with ice-penetrating radar instruments and multi-beam lasers across its larger sibling in East Antarctica.

By combining all the data they receive from satellite surveys, the POLENET sensors and radar measurements, the researchers are now seeing a clear trend. Not only is Antarctica losing land ice from both of its ice sheets, the ice loss is also accelerating.

The world in a bubble

Like the sediments on the seafloor, Antarctica's ice itself is also a collector of information from the past, albeit a more myopic one. Data from ice cores stretch back only tens to hundreds of thousands of years, as opposed to the many millions of years covered by sediment layers, but their analysis has had a particularly significant impact on climate science because it has demonstrated how closely temperature and greenhouse gas concentrations were linked in the past.

Perhaps most famously, one of the oldest ice cores was extracted at the Russian Vostok station in East Antarctica where the ice sheet is almost 4000 metres (2.5 miles) thick. The effort yielded a remarkably deep column, down to 3623 metres (11,886 feet, 3962 yards), with ice that had been deposited 460,000 years ago, well before Neanderthals and *Homo sapiens* evolved. For a long time, the Vostok ice core set the benchmark for discussions about modern climate change. Today, it remains the longest in terms of depth, but the ante was upped in 2005 by a multi-national European project, which chose Dome C, a site almost 600 kilometres (370 miles) from Vostok, to extract an unbroken sequence of frozen layers reaching almost twice as far back into the past to 800,000 years ago.

Antarctica's ice cap is crowned by seven flat snowy summits, and Dome C is one of the highest. It is also one of the coldest and loneliest places on Earth. Temperatures hardly rise above minus 25 degrees Celsius (-13 °F) in summer and can fall below minus 80 degrees (-112 °F) in winter. Snow stretches to the horizon in all directions, and the occasional skua that has learned to fly across the continent, rather than circumnavigate it, is the only sign of life. The climate record from the Dome C ice core reveals eight previous glacial cycles but, perhaps more importantly, the climate signals match those from the Vostok core as well as those retrieved from

Glaciologist Nancy Bertler (above right) and her team at the Victoria Lower Glacier work in face masks and protective gear to avoid contaminating the ice cores.

Dome Fuji on the opposite end of the East Antarctic Plateau – which suggests that the largest part of inland Antarctica behaves as one when it comes to changes in climate.

One crisp morning during my first Antarctic visit, I woke up to an urgent but welcome message: if I wanted to be on the next helicopter flight heading out to the ice-core team, I had half an hour to get ready. Moments later I was dragging my bags to the helicopter pad and we soon took off across the still-frozen McMurdo Sound, heading towards the Dry Valleys and the Victoria Lower Glacier. The pilot landed just long enough for his co-pilot to offload a tent, food box, survival gear and me, and for a moment I thought I had been abandoned in the middle of white nothingness. Then I spotted what looked like a tall mast, with a person crouched at its base. Next, a few more figures began emerging from underneath the ice, dressed in white overalls, complete with hoods, and wearing white surgical face masks. Nancy Bertler's glaciology field camp was to become home for the next few days.

Bertler's project is the New Zealand part of the International Trans-Antarctic Scientific Expedition, an initiative which uses ice-core records to gauge Antarctica's contribution to global climate and to figure out how parts of the continent itself react to any climatic fluctuations. While most of the other 22 countries involved despatch their glaciologists to the interior of the continent to drill deep into the thicker and older parts of Antarctica's ice cover, New Zealand has found its own scientific niche at low-elevation coastal sites. The information stored in ice cores from such sites may only reach back a few tens of thousands of years – much shorter than cores from further inland – but the coastal regions generally receive more snow and stormy weather and are more sensitive to climate changes.

To get onto Bertler's team, a good pair of biceps and a strong back are prerequisites – almost as important as an understanding of the physics and chemistry of glacial ice. From the air, the only visible parts of her camp are the tents and drill rig, but underneath the small drilling platform is a massive trench, excavated by

ABOVE LEFT Each season, Bertler's team has to excavate a massive cavern below the drilling platform that acts as a temperature-controlled laboratory.

ABOVE RIGHT Alex Pyne is the engineering mastermind behind New Zealand's ice and sediment drilling projects. ANTARCTICA NZ PICTORIAL COLLECTION: TAMSIN FALCONER/K001 05/06 hand each season, which serves as a temperature-controlled laboratory. When I arrived, the team had already put in almost six weeks of fieldwork, drilling hundreds of yards deep into the glacier to collect samples of ancient ice. About every half-hour, a metal pipe slid down a chute into the underground cavern and onto an impromptu bench made from white containers. After a quick inspection and some measurements, the translucent blue core within was wrapped up, labelled and stored in shelves that had been cut into the ice along the walls with a handsaw.

Above, on the surface of the glacier, Alex Pyne was in charge of the drill rig. As the engineering mastermind behind all sediment- and ice-drilling operations run by New Zealand since the 1980s, he was soon to become a familiar sight. He had started out as a geology student in Peter Barrett's group, but his extraordinary talent for solving technical challenges and developing new drilling techniques, especially for Antarctic conditions, was soon recognised. His input has been pivotal to the success of both the Cape Roberts and ANDRILL projects, and he is an equally essential member of Bertler's team, with an almost intuitive sense of whether the drill bit hundreds of yards below is doing what it is meant to do.

A few Antarctic summers later, I found Bertler back at Scott Base, waiting for a break in the weather to get her team and gear up to the Mount Erebus saddle. The area is notorious for its foul weather but that is exactly why this site is so important to her. Antarctica's coast is exposed to a storm belt that circles around the continent and it experiences larger changes in atmospheric circulation than the interior. Further inland, the record of such changes is lost simply because of the inertia of the massive ice sheets. Compared to the continent's coastal areas, inland Antarctica is a calm place, with a climate driven by very cold air from high up in the stratosphere, 10 to 50 kilometres (6 to 30 miles) above the Earth's surface. The amount of snowfall along the coastline can be many times that in the interior, which means that the seasonal layers of ice are thicker and more easily resolved, like annual tree rings. The Mount Erebus saddle is one of a handful of coastal sites with extremely high annual snowfall and, once Bertler managed to get there, it delivered ice cores with some of the best resolution and exceptional climate sensitivity.

For Bertler, each ice core is a treasure trove. The frozen water, dust particles and bubbles of ancient air trapped in the ice each reveal precious information about past atmospheric conditions. From the water and its isotope ratios, she can deduce the temperature at the time the snow fell and whether the air mass that precipitated it came from the tropics or from the Ross Sea region, indicating changes in atmospheric circulation. The water also contains chemical markers of certain algae which help her trace the extent of seasonal sea ice and associated blooms of

Nancy Bertler, right, and her team in their icy makeshift laboratory on the Mount Erebus saddle.
ROB McPHAIL

plankton, diatoms in particular. Armed with these proxies, she can estimate how much of the sun's heat was reflected back into space by sea ice rather than being absorbed by the ocean.

As snow falls, it takes with it aerosol particles and some gases that have an affinity for water. It is in this part of the record, in the materials trapped in surface snow, that Bertler can identify spikes of sulphuric acid from major volcanic eruptions, as well as components of sea salt and terrestrial dust. The dust that washes out from the atmosphere tells her something about the frequency and intensity of storms and how vigorous the atmosphere was in the past. Its chemical composition also reveals where the dust came from and, perhaps surprisingly, it shows that the biggest source of dust in Antarctica is South America rather than Australia or New Zealand.

However, there is one thing only ice cores can provide. Frozen in tiny bubbles inside the ice are samples of ancient atmospheres – tiny windows onto our ancestral world. The air inside these bubbles still includes the same mixture of elements it had at the time, including greenhouse gases such as carbon dioxide and methane, and provides a snapshot of past climate conditions. The high snow accumulation rate and strong winds at the Mount Erebus saddle help to trap these air bubbles in the ice faster than in other parts of the continent, which gives Bertler a good chance to trace changes in greenhouse gas concentrations on an annual basis and to study the causes of any significant shifts.

The future challenge for glaciologists worldwide is to start drilling in some of the most inaccessible parts of the frozen continent. Dome A is the tallest of the summits on the eastern ice sheet and the last to be conquered. The first people to step onto it in 2005 were part of a Chinese team that had trekked 1200 kilometres (750 miles) from the coastal station of Zhongshan. The scientific importance of Dome A is not its height itself but the ice below. Three kilometres (1.9 miles) beneath the peak may well be the world's oldest ice. The rationale for this treasure hunt is that the older ice would push the climate record to one of the more interesting times in Earth's glacial history when, about a million years ago, the rhythm of glaciations began to change from Milankovitch's 41,000-year cycles to 100,000-year events. Sediment cores suggest that the key transition between these states took place over a period of several hundred thousand years, and one explanation is that concentrations of atmospheric carbon dioxide dropped and cooled Earth down. The only way to find out is to drill for ice that is older than a million years and to extract a sample of the atmosphere at the time. Dome A receives even less precipitation than Dome C – and next to nothing compared with Antarctica's coastal sites – which means that the ice below the Dome A summit has formed from very old snow.

In the meantime, the analysis of the Dome C core has confirmed a strong link between greenhouse gas concentrations and temperature, supporting the idea that these gases, and carbon dioxide in particular, play a significant role in amplifying climate cycles. It has also provided glaciologists with the natural levels of greenhouse gases for the past 800,000 years, showing that modern carbon dioxide concentrations are more than 30 per cent higher than those measured in the ice cores.

While carbon dioxide, some of which remains in the atmosphere for thousands of years, proved to be an important driver for large shifts from glacial to interglacial periods, the shorter-lived methane is seen as a cause of rapid climate switches that can happen within decades. In scientific terms, these super-fast changes are referred to as Dansgaard–Oeschger, or D–O events and they are most obvious in ice cores from Greenland, which reach back to the end of the last interglacial period. But Antarctic ice cores suggest that there is a coupling between these rapid climate shifts in the two hemispheres – a seesaw with the world's oceans as a connecting bridge and the poles as the seats at each end.

The ocean superhighway

As a young graduate student at Columbia University's Lamont-Doherty Earth Observatory in New York in the 1960s, Wally Broecker wanted to study the oceans because he realised that they connected all continents. It was the start of a long career which had a profound impact on the understanding of the oceans themselves, as well as their role in global climate. Broecker developed several geochemical tracers to describe how carbon dioxide is processed in the oceans and how it flows between the oceans and the atmosphere. During natural climate cycles in the past, carbon dioxide fluctuations are viewed as an almost universally oceanic phenomenon, a consequence of the large pools of carbon sequestered there. Changes in ocean circulation, biological productivity, carbon dioxide solubility and other aspects of ocean chemistry all interact to determine whether oceans act as a sink or source of carbon.

The chemical capture or release of carbon is only one part of the equation, though. Broecker coined the term Global Ocean Conveyor Belt when he described the role of the major ocean currents in shifting heat, salt, gases and nutrients around the planet. It is via this conveyor belt – now better known as the thermohaline circulation – that Europe and eastern North America remain relatively warm and the rest of the world is more exposed to polar influences.

The idea provides a big-picture outline of the world's main ocean currents and the route they take as they travel between the poles. In this big picture, Antarctica and the Southern Ocean are major components, according to Victoria University marine geologist Lionel Carter whose research has focused on the more subtle nuances of global ocean flows. Carter's grey-haired elegance belies the fact that he has spent much of his career on board ship in some of the most turbulent seas, tracing the journey of currents between deep-ocean basins and coastlines. With the help of isotope and other forensic techniques similar to those used by geologists, it is possible to trace a body of water as it flows through the ocean like a river, and to track the movement of heat to far-flung places around the world. The journey begins in Antarctica.

Each southern winter, the surface of the ocean around Antarctica freezes, doubling the size of the continent. In spring, the ice breaks up into floes. This view across McMurdo Sound looks towards Mount Bird on the left and Mount Erebus in the centre. ROB McPHAIL

Each winter, Antarctica becomes the stage for the planet's largest seasonal display as it doubles its already vast size by forming an apron of sea ice along its coastline. From a climate point of view, one of the important consequences of this large-scale freezing of the ocean is that the ice acts as a silver screen, reflecting the sun's energy back out into space. Another important factor is that, as the ocean freezes, the ice on the surface expels salty water underneath. The brine is heavier than the underlying ocean and it begins to sink to the bottom around the margins of Antarctica. Some of it is picked up by the Antarctic Circumpolar Current, which can

only be described in superlatives. It is the longest current, with an estimated journey of 24,000 kilometres (15,000 miles). It transports the biggest volume of water and, because it connects all other oceans, it plays the most important part in the global distribution of all those ocean ingredients mentioned earlier. This current gains further momentum from the westerly winds of the circumpolar vortex, which accounts for 70 per cent of the wind energy that drives ocean currents worldwide.

For most of its passage, the Antarctic Circumpolar Current is guided along the flanks of mid-ocean ridges in water depths of around 3500 metres (2 miles),

The Antarctic Circumpolar
Current, the world's longest,
connects the oceans.
SPENCER LEVINE, AFTER
LIONEL CARTER

- - - - **ANTARCTIC CIRCLE**
        66.56°S
—— **ANTARCTIC CIRCUMPOLAR**
        CURRENT

widening as it passes the Kerguelen Plateau and constricting again around the
large submarine elevations of the Campbell and Falkland plateaux. As the current
meets these topographical features, it forms giant eddies that migrate along the
edges of the submarine plateaux, scouring the deep-ocean floor en route. At depths
below 2000 metres (1.2 miles), large currents are tapped off from the circumpolar
highway and extend north into the Atlantic, Pacific and Indian oceans. These deep
currents with their cargoes of Antarctic waters contribute to the thermohaline
circulation, which, as the name suggests, is driven by differences in temperature
(thermo) and salinity (haline). These factors determine the density of sea water
and, along with the wind and eddies, keep the global circulation on the move.

Lionel Carter on board
the NIWA research vessel
Tangaroa in 2002. Using
ocean temperatures and
salt concentrations, as
well as satellite data,
he can trace a current
as it flows through
the ocean like a river.
COURTESY OF LIONEL CARTER

The two big engine rooms for the production of cold and salty water are near the poles, around Antarctica and in the North Atlantic, from where the water flows to and through deep-ocean basins. This movement is compensated by the upwelling of shallower waters to the surface in the equatorial regions and their return to polar environs as warm and deeper bodies of water. Without the two polar engines, such warmer currents would never reach the coastlines of Europe, America and most places where people have settled. The Gulf Stream is probably the best known of these warm surface currents which are critical for bringing milder temperatures to some of the most densely inhabited areas.

As Tim Naish puts it, messing with the planet's engine rooms could have dire consequences. Earth's largest freshwater reservoir sits on top of the Antarctic continent and any melting of its ice sheets would flood the Southern Ocean with salt-free water, which in turn would upset the density structure of the ocean, with the potential for slowing or even shutting down the thermohaline circulation. While he is not suggesting that this is going to happen soon, it has happened in the past, and sometimes very quickly. Massive flows of fresh water from lakes in North America, combined with deglaciation, are thought to have led to exactly such a disruption in the production of cold and deep water in the North Atlantic engine room and caused a mini-ice age in Europe known as the Younger Dryas.

In 2009, three hardy souls – University of Alaska physicist Andy Mahoney, University of Otago PhD student Alex Gough and Brian Staite from Antarctica New Zealand – spent the winter at Scott Base to gather the first year-round observations of sea-ice formation and thickness in McMurdo Sound. Sea ice forms when the ocean around Antarctica begins to freeze at the start of winter. While satellite images give a good indication of the extent of the freezing, they don't show how thick the ice is or how it forms around ice shelves that extend from almost half of the continent's coastline. The team travelled out onto the sea ice for up to three days at a time, working in total darkness and frigid temperatures. They carried out two main experiments: measuring thickness and crystal structure of multi-year sea ice at Cape Armitage and of first-year sea ice at Erebus Bay.
This image shows Alex Gough installing a string of sensors to record sea ice temperature.
ANDY MAHONEY

Also known as the Big Freeze, this period was brief, geologically speaking, lasting about 1400 years after its abrupt onset 12,900 years ago. Within a decade or so, the average annual temperature in Europe dropped by about five degrees Celsius (9 degrees in Fahrenheit) and much of the landscape disappeared under ice fields and glaciers.

Nothing quite this fast and extreme has been experienced since. But one of the fundamental conclusions from the study of ancient climate conditions is that today's concentrations of greenhouse gases have no parallel in at least the last 800,000 years, the time span measured by ice cores, and that when greenhouse gas concentrations were as high or higher than today in Earth's more distant past, the world was a very different place indeed. It seems that our planet's climate responds if pushed, but that it doesn't always respond smoothly but rather in shudders and sudden jumps.

For Peter Barrett, the prospect of significant climate change through anthropogenic emissions of greenhouse gases is now one of the main motivators for his research and the changes he is making to reduce his own carbon footprint. He sees the present climate in Antarctica as a response to slow changes that happened on a timescale of hundreds of thousands of years. Yet projections for 2100 suggest that global temperatures could rise by between two and five degrees, depending on the extent to which greenhouse gas emissions can be reduced in the meantime. Satellite monitoring is already showing that Antarctica is losing ice. Although the rate is small, it is accelerating and Barrett feels there is no time to lose.

John Mercer's prediction of disappearing ice shelves became dramatic reality when the largest section of the Larsen Ice Shelf collapsed in January 1995, followed by Larsen B, the next biggest segment, in February 2002. Some of the huge ice masses these shelves once held in place are now flowing faster through the scars left behind by their collapse. This is why, when it comes to global warming, Barrett doesn't mince words. He caused a stir when he predicted that we are facing the end of civilisation as we know it by the end of this century, unless there is determined action to curb greenhouse gas emissions. But he stands by that statement, warning that even if temperatures remain in the lower part of the predicted range, an increase in storm power and the extremes of floods and droughts will profoundly affect people and their livelihoods right around the globe. If temperatures are in the upper part of that range, he says, Earth will return to the climate it once had before the big ice sheets formed in Antarctica 34 million years ago – a long time before humans, or in fact any of our close evolutionary ancestors, appeared. Antarctica, hostile and unforgiving to all but the hardiest survivors, could once again become a temperate oasis at the edge of an increasingly hot and once civilised world.

Thinning ozone shield

The crew of the *Belgica*, a whaling ship that had been refitted for a Belgian expedition in 1897, were the first people to live through an Antarctic winter. It wasn't part of the expedition's original plan, but the ship was caught in the pack ice in early March of the following year, and the men, Norwegian Roald Amundsen among them, had no option but to wait for the ice to melt again in spring.

They were ill-prepared for the long polar night. Short of food and clothing, many expedition members became incapacitated by scurvy or came close to losing their sanity during the gruelling months of darkness and freezing cold. Nevertheless, they managed to continue gathering meteorological observations throughout winter. This was an Antarctic expedition with a solely scientific mission, and when the ship eventually returned to Belgium in 1899 the men were treated to a hero's welcome and celebrated for bringing back the first continuous record of weather data for a full Antarctic year.

Unbeknown to them, they were also the first to see polar stratospheric clouds, unusual formations that look like giant, airborne mother-of-pearl shells. The clouds are so high up in the atmosphere that they can capture and reflect the sun's light from over the horizon. The *Belgica* crew had no way of knowing that these strange, luminous clouds would in time help explain the perplexing phenomenon of ozone depletion.

Mario Molina was a young scientist at the University of California, Irvine, during the early 1970s, when he made an unsettling discovery. He was interested in a group of chemicals that were used widely in fridges and spray cans. Chlorofluorocarbons, or CFCs, were thought to be chemically inert, but Molina was investigating what would happen once they reached the upper atmosphere.

James Lovelock, who would soon become famous for his Gaia theory of Earth as a self-regulating complex system, had developed a highly sensitive device for measuring extremely low concentrations of gases high above the Earth's surface, and he showed that CFCs had already spread globally throughout the atmosphere. Molina realised that far from being inert, CFCs had the potential to destroy ozone under certain conditions that exist in the stratosphere – the upper part of the atmosphere between 10 to 50 kilometres (6 to 30 miles) above the surface. There, a thin layer of ozone, a molecule of three oxygen atoms, shields the planet from the sun's intense ultraviolet radiation. Molina feared that if his predictions were right, the world would be in trouble.

In 1974, Molina and his supervisor Sherwood Rowland published a paper warning that CFCs could be transported up to the ozone layer, where ultraviolet radiation would split them into their constituents, notably chlorine atoms. They calculated that if human use of CFCs continued unabated, the increasing number of free chlorine atoms in the stratosphere would significantly deplete the ozone layer and lead to a dangerous increase in ultraviolet radiation at the Earth's surface. Their work led to certain restrictions on CFCs during the late 1970s, but the real urgency of the problem became clear a decade later when Joseph Farman, Brian Gardiner and Jonathan Shanklin, all three from the British Antarctic Survey, published their observations from Antarctica. Their data showed that ozone levels above the frozen continent had started dropping dramatically during each southern spring.

The first ozone measurements in Antarctica had been taken by a British team in 1956, during preparations for the International Geophysical Year, as part of a larger, long-term project to begin systematic meteorological observations from the ice. Satellite measurements followed in the 1970s. By the time Farman and his colleagues published their observations in 1985, they had more than twenty years of data which showed that ozone levels had begun to decrease by the early 1980s, and an extraordinary change happened each spring when a 'hole' opened up in the ozone layer above the continent. The report shocked the world, but some scientists continued to question the validity of the data. Susan Solomon was one who took it seriously from the start.

A young chemist working for the National Oceanic and Atmospheric Administration in Boulder, Colorado, Solomon set out to explore what could be causing such dramatic depletion in the ozone layer above the bottom of the world. As she was considering various ideas that were being put forward, she kept coming back to chlorine as an important player in the high-altitude chemistry of ozone, but was intrigued as to why it should cause so much more damage above Antarctica than anywhere else. The rapid depletion of Antarctic ozone could not be explained by conventional chemical reactions between gases. There had to be another important mechanism that helped speed up the process in Antarctica. Solomon's answer to the mystery was polar stratospheric clouds.

Circular wind patterns isolate the air above Antarctica from its surroundings, and during the polar winter, when temperatures in the stratosphere above the continent hover below minus 80 degrees Celsius (-112 °F), clouds of water and nitric acid condense in this high-altitude vortex of cold air. The icy particles in these clouds provide the surface for some of the chemical reactions that begin to destroy ozone as soon as the light returns to the frozen continent in spring.

When Solomon wrote a few radical papers suggesting that the clouds were an essential element in the process of rapid ozone depletion, her work was dismissed at first, just as the reports of an ozone hole above Antarctica had been discounted earlier. But a year later, in 1986, others had confirmed the ozone hole data and Solomon found herself preparing for a winter expedition to Antarctica to observe first-hand what exactly happened as the ozone shield above opened up.

Solomon's team used spectroscopy, measuring the absorption of certain wavelengths of light from the moon and the sun, to determine the concentrations of ozone and various other chemicals, in particular chlorine dioxide. Other teams used different methods to monitor the chemical processes high above them, and collectively they produced a comprehensive picture of the dramatic events during the Antarctic spring. Proof was provided that chlorine compounds were indeed enhanced in the stratosphere, by a factor of 100 in the case of chlorine dioxide, compared to what concentrations should have been without surface chemistry. They also showed that polar stratospheric clouds influenced chemical processes in ways that were consistent with the observations of an ozone hole.

The findings were still not enough to clinch the debate, but a year later, when an aircraft mission completed another set of measurements by yet another technique, the lines of evidence were strong enough to prompt international action. The Montreal Protocol came into force in 1989, under the Vienna Convention for the Protection of the Ozone Layer, designed to phase out the production of ozone-depleting compounds. In 1996, Molina and Rowland shared the Nobel Prize in Chemistry, together with Paul Crutzen who had led equally relevant work about the role of nitrogen oxides in ozone chemistry.

Tracking the hole

Arrival Heights is a small, dome-covered laboratory overlooking McMurdo Station. It was built in 1960 and its original function was as a radar station to record auroras. Since then, it has filled up with various instruments, peeking out from underneath the revolving dome, that remotely monitor the chemistry of the atmosphere. The dome is guided by sun trackers to make sure that all detectors are always pointed directly into the light source.

When I visited in 2001, I was particularly keen to see the Dobson spectrophotometer, one of five such instruments in Antarctica and one of 150 in a global network that measures ozone every day to monitor the ozone layer and to ground-proof (or spot-check) satellite data. Developed in 1926, it is the earliest instrument used to measure ozone, and each of the small unremarkable boxes worldwide can trace its pedigree directly to its inventor, British physicist Gordon Dobson.

Continuous ozone measurements at Arrival Heights started in 1988, just before the Montreal Protocol came into force, and for much of the time since, scientists at New Zealand's National Institute of Water and Atmospheric Research (NIWA) have been in charge of the black box they simply call the Dobson. As they watched the annual ozone holes expand and shrink, and then dissipate again in summer as the polar stratospheric clouds dissolve in the warming air, they helped calculate trends and projections for the future. In 2008, two decades after the ban of CFCs,

This antenna at Arrival Heights was set up by University of Otago physicist Craig Rodger to track changes in solar energy with the help of very low frequency (VLF) radio waves (5 to 30 kilohertz), with wavelengths of about 10 to 60 kilometres (6 to 37 miles). Such waves are either broadcast by large communications transmitters or produced by lightning, and the antenna contributes to two global monitoring networks – one that tracks changes in the sun and one that observes lightning storms.

Rodger is interested in particles emitted by solar flares, which are ejected into space with enormous force. If the explosions are directed at Earth, they can trigger geomagnetic storms in the ionosphere, the charged part of the atmosphere about 70 to 90 kilometres (45 to 55 miles) above the Earth's surface. Low-level explosions cause storms that result in auroras, but more energetic explosions emit particles that can change the chemistry of the atmosphere and contribute to ozone depletion, hence the need for monitoring.

While everybody is aware of the ozone hole high up in the stratosphere, few people realise that sudden and rapid ozone depletion also happens in the lowest part of the polar atmosphere, in the marine boundary layer. These springtime ozone depletion events can happen in just a few hours, with ozone concentrations dropping from normal levels of 25–40 parts per billion (ppb) to as low as 0.05 ppb. The small amount of ozone this close to the ground has no significance for filtering harmful ultraviolet rays, but NIWA scientists are keen to understand the processes involved in its depletion

The main mechanism for this sudden loss of ozone involves the release of bromine compounds from sea salt, in what's known as a 'bromine explosion'. NIWA scientists Karin Kreher and Paul Johnston have been measuring bromine oxide since 1995 and surface ozone in collaboration with US colleagues since 1997 at the Arrival Heights observatory near Scott Base. In 2006, to measure bromine explosions at more remote sea-ice sites for the first time, the NIWA team deployed a portable instrument, towing the insulated, waterproof system out to several sites on McMurdo Sound (above) and Cape Bird. Below NIWA scientist Katja Riedel stands with the instrument on the beach at Cape Bird. TIM HAY

scientists published the first cautious notes that the ozone layer may have begun its recovery, predicting that it could return to normal levels not long after 2050. There is one complicating factor, though. Ozone depletion and climate change influence each other.

By the time I first met Susan Solomon in 2001, during a writer's tour to promote her book about Scott's last journey, an Antarctic glacier had been named after her and she was about to embark on her next big scientific challenge as co-chair of a working group for the Intergovernmental Panel on Climate Change. Scientists had been reluctant to bring ozone depletion and climate change together to avoid confusion between the two processes, but a growing number of mutual feedbacks between them are becoming increasingly clear.

Ozone, CFCs and the substances that have been introduced as replacements are all greenhouse gases and are contributing to climate change. In turn, the warming of the Earth's surface goes hand in hand with a cooling of the higher atmosphere, which will likely increase Antarctic ozone depletion and may slow down the recovery of the hole. However, climate change will also affect how the cold-air vortex above Antarctica forms and the chemistry that occurs within it, along with the concentrations of other greenhouse gases, some of which could slightly accelerate the recovery of the ozone layer. The collective net effect at this stage is uncertain.

Compared to climate change, Solomon looks back at the ozone issue as an easy one to explain and take action upon. There was one dramatic effect that could be measured in one season, it wasn't too difficult for industry to come up with substitutes for CFCs, and all the products that once relied on them are still in use today. There is no equivalent to these solutions in the area of climate change, yet Solomon sees the evidence for such change as compelling, and its consequences as far more pernicious and long-lived.

In the case of ozone depletion, Antarctica's ozone hole was the most dramatic manifestation of a chemical process that was happening elsewhere as well, thinning the ozone shield and reducing its protection against ultraviolet rays above some of the most heavily populated areas of the planet. With climate change, the polar regions are also reacting profoundly and quickly, like remote warning posts. But their role is even more complex because they not only respond to but also drive some of the climatic effects.

2
Life on Ice

The emperors inspect their own stylised form in Antarctica
New Zealand's logo on the helicopter. ROB McPHAIL

On all sides rose great walls of battered ice with steep
snow-slopes in the middle, where we slithered about and
blundered into crevasses. To the left rose the huge cliff of
Cape Crozier, but we could not tell whether there were not
two or three pressure ridges between us and it, and though
we tried at least four ways, there was no possibility of getting
forward. And then we heard the Emperors calling.
– Apsley Cherry-Garrard, *The Worst Journey in the World*

Short of retracing Apsley Cherry-Garrard's *Worst Journey in the World*, midwinter
and in starlit darkness, one has to be lucky to see emperor penguins. I was, and the
encounter is one of my most exquisite Antarctic memories.

Antarctica is an unforgiving frigid desert surrounded by a frozen ocean. Any
creature that manages to survive on land or in the water is operating at the limits
of biology, and the freezing temperature is only one of many challenges to life. It is
completely dark for four to six months of the year and food supplies fluctuate from
feast to famine with the severe changes between seasons. Fresh water is abundant
but inaccessibly locked up in ice. Any attempts to breed and raise a brood can be
life-threatening endeavours for both generations. A small mistake or deviation
from a finely tuned survival tactic is usually fatal.

Emperor penguins are champions of Antarctic survival. They may well be the
only birds never to set foot on land. They forage from floating chunks of ice and
breed on thick slabs of sea ice, in the depths of winter. They are so well adapted to
their frozen home that they could not survive without ice. More than any other
Antarctic wildlife, emperors have become a symbol of endurance and resilience in
the harshest conditions.

My encounter happened at the end of a long day and at the fringe of the continent. With 24 hours of daylight during summer, everyone puts in long shifts. I
had already made a bone-rattling Hagglund journey across the sea ice to Captain
Scott's Terra Nova hut to meet a team of conservators. Then came a helicopter
flight to the Victoria Valley to join geologists studying the strange polygonal shapes
that form almost everywhere in the hyper-arid McMurdo Dry Valleys, followed by
a skip across to the Taylor Valley and the permanently ice-covered Lake Fryxell to
catch a group of biologists as they emerged from a dive to the murky lake bottom in
search of life, microbial or otherwise.

Sitting still was the best strategy for getting a close view of the emperor penguins. Though they kept at a safe distance, they came closer to us than we would have been allowed to approach them. Scott Base chef Donna Wightman, right, and Antarctic Field Training instructor Abel Roche, centre, are closely observing with me.
ROB McPHAIL

I wasn't surprised when our helicopter pilot decided to take the long route back to Scott Base. Rob McPhail has clocked up thousands of flying hours during more than twenty consecutive Antarctic summers and he knows this part of the continent like the back of his hand. Numerous scientists have relied on him to get to their field camps and back safely and, by way of thanks, lobbied to name a dramatic ridge of wind-sculpted granite towers at the entrance to the Wright Valley the McPhail Turrets.

During our return flight, everybody felt absolutely safe in his hands – safe enough to nod off despite the spectacular scenery. I only noticed a hint of concern when I realised he was flying midway across the frozen ocean towards the edge of the sea ice, and then dropping down to land. What followed was two hours of awed and blissful silence. There was no need to say much as we walked slowly along the ragged, frozen and ever-changing edge of Antarctica, under the biggest blue skies, surrounded by a group of curious emperors, frolicking in the ocean and jumping onto the ice to inspect us.

Emperors are the giants of the penguin world, easily eye to eye with a person squatting on the ice. The playful young birds we encountered were not ready to breed yet but a few seasons later they would join others in an epic and stoic effort to perpetuate their species. Their midwinter breeding cycle seems an almost suicidal mission, but biologists think that the strategy evolved to give fledgling emperors the best survival chances as they grow to independence at a time when the ocean opens up and food is plentiful.

To start the annual breeding routine, thousands of adults have to trudge across more than 100 kilometres (60 miles) of rugged sea ice just to mate. They choose a site where the ice is stable, either near an ice shelf or in the lee of an island. All birds stay until the eggs are laid six weeks later. Then, the females trek back across the widening ice to feed at sea. The males stay put, incubating the egg between their feet and a roll of skin and feathers. By the time they get this far, there seems little point in building nests or defending territories. With temperatures down to minus 60 degrees Celsius (-76 °F), chilled even further by frequent icy blizzards, it makes

A waterproof coat of feathers and a thick layer of blubber protect emperors and other penguins from the cold. ROB McPHAIL

A lone orca hunting at
Cape Bird. ROB McPHAIL

Penguins and chicks
covered in ice and snow.
ANTARCTICA NZ PICTORIAL
COLLECTION: RACHEL BROWN,
GUS McALLISTER/K002 05/06

more sense to cast aside any territorial aggression and huddle together as closely as possible. An individual penguin would have a slim chance of survival, but as a collective, with up to 6000 birds wrapped around each other in a huge crowd, the colony acts almost like a separate organism in its own right. As each egg-hogging male warms up in the centre of the huddle, he slowly shuffles back to the edge of the colony to take the brunt of the cold for a while, all the while balancing his precious cargo in his portable nest and going hungry. The eggs can be an astonishing 70 to 80 degrees (120 to 140 degrees in Fahrenheit) warmer than the surrounding air.

By the time the females return nine weeks later to take a turn in parenting once the chick hatches and is ready to leave its brood pouch, the males have lost nearly half their body weight and are desperate to get back to the ocean to feed. As if their effort so far hadn't been enough, they return to take turns in feeding their growing chick. As the hatchlings grow and spend longer spells without either parent, they instinctively begin to practise the life-saving huddling in downy crèches.

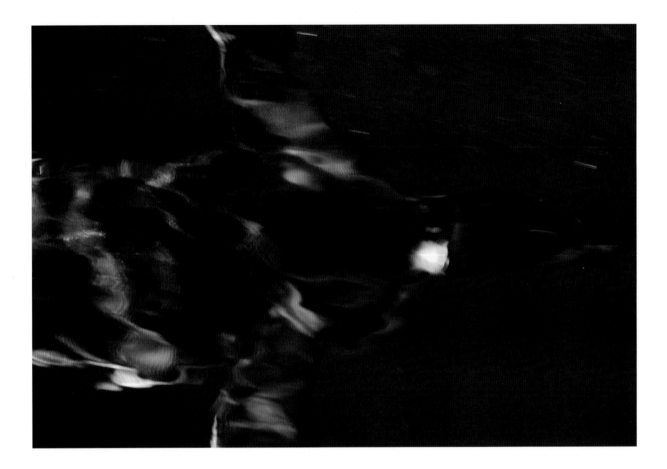

Emperor penguins can dive deeper than 500 metres (0.3 miles) and stay under water for up to 22 minutes.
ROB McPHAIL

They eventually shed their soft down at the height of the Antarctic summer, but when they enter the ocean they weigh only about 60 per cent of an adult's weight, the lowest value for any penguin. During harsh years only one in five survives the first year, but if they make it to the next summer, their chances are good. With a nine-month breeding cycle, the adults only have a few weeks left to fatten up again, moult and fatten up some more. Despite the exertion, adult emperors have remarkably high survival rates compared to other penguins, with an average of 95 per cent returning to the breeding site to do it all again next winter.

The emperors' extraordinary behaviour is only one of many adaptations that allow the birds to survive the extreme conditions of their icy home. Wrapped up in a waterproof coat of feathers and a thick layer of blubber, they are superbly insulated. Their size is driven by the physics of heat loss, which predicts that bigger is better when it comes to keeping warm. The only body parts that do cool down are their feet and flippers, and to cope with the temperature gradient between their body and its extremities, emperors and other penguins have changed their anatomy and evolved a counter-current blood-heating mechanism. Their veins and arteries run much closer together than in other animals to ensure that the cold blood returning to the body is heated up along the way. The muscles that operate an emperor's feet and flippers lie deep within the warm body, attached to long tendons that act as remote controls, transferring movement at a distance. When it gets extremely cold, the birds simply slump down to cover their feet with feathers and body blubber to avoid wind chill and frostbite.

Ghosts of the ocean

All life in and around Antarctica has evolved within very narrow margins, calibrated to cope with the extremes and to make the best use of the scarce resources when they are available. The evolutionary forces at play are so powerful they have produced animals and plants that look and behave like no others. While images of ice-encrusted downy penguin chicks huddling together for warmth are usually the first to come to mind, in terms of extreme adaptation, Antarctic fish truly stand out.

The men on board HMS *Terror* were most likely too exhausted to appreciate their extraordinary discovery. It was the end of February 1842, and they had been sailing south from Tasmania for weeks on one of the final legs of a four-year expedition that had begun in 1839. Theirs was one of the last major voyages made entirely under sail, led by polar explorer James Clark Ross.

A southerly wind had built up to a violent gale. Waves breaking over the ship froze almost instantly as they fell on the deck and rigging, and ice was building up to a dangerously heavy load on the hull and ropes. Crew members were chopping away at the thick coat of ice with axes, trying to clear the bows of their ship and the expedition's second vessel HMS *Erebus*, when a small fish emerged from the blocks. It looked like nothing any of them had ever seen before: pale and translucent, with big eyes and a spoon-shaped toothy jaw. Joseph Dalton Hooker, the expedition's assistant surgeon and naturalist who would eventually rise to fame in Britain and beyond, had just enough time to make a quick sketch of the peculiar creature before the ship's cat gulped it down.

It would be nearly a century before anybody was able to take a closer look at Antarctica's unique icefish. Norwegian biologists on an expedition to the subantarctic Bouvet Island in 1928 caught a specimen and found that the fish was indeed transparent because its blood was white. Later research confirmed that it had no red blood cells, the pigmented oxygen-carrying cells found in every other living vertebrate. Another four decades later, the first molecular analysis of bodily fluids of icefish showed that they had not only lost the cells but the pigment – haemoglobin – as well. One of the genes responsible for making the large haemoglobin protein was a molecular fossil, a remnant still present in the fish DNA but utterly useless without the second gene, which had gone completely extinct. The fish had abandoned the ability to make a molecule that has been crucial to the lives of their ancestors for over 500 million years. The clear fluid running through their veins was not much different from icy water, with just 1 per cent of the volume made up of white blood cells.

Until the discovery of icefish, most biologists thought the sub-zero (sub-32 °F) waters of the Southern Ocean were too cold to support any fish fauna. Yet most Antarctic expeditions during the nineteenth and twentieth centuries returned with more unusual creatures, until eventually more than 200 different fish had been described. Even today, biologists are still discovering new fish species in the Southern Ocean. Most of them are part of one highly adapted group that dominates the seas around the frozen continent and is found nowhere else in the world, giving Antarctic fish an extraordinary degree of endemism.

The fifteen or so species of icefish remain the only ones to have discarded haemoglobin production entirely, but all other Antarctic fish have dropped the number of red blood cells per volume to about a third of that found in other vertebrates, including humans. They simply can't afford any more cells. In the sub-freezing temperatures of the Antarctic ocean, their blood would become too thick and difficult to pump, like engine oil during a cold winter. On the upside, oxygen is more soluble in cold water, and instead of relying on haemoglobin to bind and transport it via the bloodstream, Antarctic fish absorb oxygen directly through their skin and enlarged gills.

The loss of a function as vital as oxygen transport might seem extreme, and in warmer waters it would have been fatal. However, in the cold oceans around Antarctica it was tolerable and eventually turned out to be an advantage, but it would not have been enough to survive. Survival required invention. Some 25 million years ago, when the opening of the Drake Passage completed Antarctica's isolation and established the Southern Ocean, the continent continued to cool. As the growing ice sheet clawed into the shoreline, habitat disappeared and the existing coastal fish fauna died out almost totally. There is no fossil record of modern Antarctic fish. Neither are there any representative species or relatives in the Northern Hemisphere and therefore biologists believe that the original fish fauna was replaced by a proliferation of new cold-adapted forms, which evolved from an ancient stock of slow-moving bottom-dwellers.

These ancestors spent their lives roaming shallow coastal waters or the muddy seabed. They had no need for a swim bladder – an internal gas-filled sack that helps fish control their buoyancy and stay afloat at a certain water depth without having to exert extra energy swimming. Like the wings of birds that no longer need to fly, the swim bladder in this group of fish eventually disappeared.

Once they became trapped by the current that circles around Antarctica, with little competition or predation, these ancestral fish quickly evolved to fill the newly vacated ecological niches. Like Darwin's finches, they swarmed out into every available living space, becoming increasingly specialised and separate.

The Antarctic shelf drops to about 500 metres (1600 feet), with depressions down to a kilometre (3200 feet). The weight of the continent's icy blanket pushes it further down than temperate shelves, which are usually no deeper than 200 metres (660 feet). Unlike temperate fish, many of Antarctica's species prefer the deeper waters, roaming anywhere between 300 to 600 metres (1000 to 2000 feet) in depth. Most remain true to their ancient bottom-dwelling lifestyle, but some took up proper swimming again, despite the lack of a swim bladder. To regain buoyancy they simply shifted the balance between the heavy and light parts of their skeleton. They lost most of their scales, replaced bone with cartilage wherever possible, and laid down fat deposits in their muscles and under their skin. Collectively, these adaptations are as good as a swim bladder and make the fish almost weightless in water, like astronauts floating in space.

Scientifically, this group of fish is known as notothenioids, or Antarctic perches, and they dominate the seas around the continent. Unlike the diverse fish fauna found in the shelf waters of other continents, almost half of all fish in Antarctic waters belong to this single suborder and they make up 95 per cent of the biomass. Within this group, the most extensively studied family, the notothens, is also the most diverse, ranging from the 60-kilogram (130-pound) Antarctic toothfish, a large mid-water predator and sister species of the Patagonian toothfish, to some of the smallest fish found in the Southern Ocean. Regardless of their size or the niche they would eventually occupy, the one major obstacle all these fish had to overcome – and the primary evolutionary leap they had to take – was to find a way of not freezing solid in the sub-zero temperatures.

Supercool lifestyle

McMurdo Sound, one of the many coastal features discovered and named by James Clark Ross during his long voyage, is one of the world's coldest bodies of water. For much of the year the sound is frozen over with more than 3 metres (10 feet, 3.3 yards) of sea ice, polished smooth by bitter katabatic winds that spill down from the Polar Plateau. Salt lowers the freezing point of the ocean below that of fresh water and, under the ice, the annual mean temperature is minus 1.9 degrees Celsius (28.6 °F). But McMurdo Sound butts up against the McMurdo Ice Shelf, an extension of the Ross Ice Shelf, the world's largest floating slab of ice, which can be up to a kilometre (0.6 miles) thick in some parts. As the ocean circulates under the ice, its freezing point is depressed even further by the pressure, and at the underside of the shelf it can be as cold as minus 2.3 degrees (27.9 °F). Still, some species of Antarctic fish that live right under the ice survive.

How they do it is a question that has puzzled Art DeVries, a taciturn fish physiologist at the University of Illinois at Urbana-Champaign, ever since his first encounter with Antarctic fish in 1961. In those days, he was working as a technician with Stanford University fish ecologist Donald Wohlschlag, who was pioneering research into how fish adapt to extreme cold. DeVries' job was to catch fish and to measure their oxygen consumption and metabolic rate. Like any good fisherman, he also spent many hours just observing – watching some fish swim among ice crystals, unharmed, while other species stayed well clear of ice in deeper water.

He was hooked, both on fish and Antarctica, after his first year on ice. DeVries had grown up in Montana and extreme cold was an attraction rather than a deterrent. Life on ice, with thawed ice for drinking water, no more than one quick shower a week and simple metal barrels known as 'honey buckets' for toilets, promised an adventure. He returned for a summer season and then, as a doctoral student, for another year, at the end of which he had almost clinched a long-standing debate about how Antarctic fish avoid freezing.

DeVries knew that fish from temperate waters would snap freeze on contact with ice crystals, so he started looking for differences in the blood of different species. As salt lowers the freezing point of water, many scientists thought it did the same in fish blood. But the blood of bony fishes is more dilute than sea water – reflecting their early evolution in fresh or brackish waters – and without protection, Antarctic fish should freeze at minus 1 degree Celsius (30.2 °F).

When DeVries found similar salt concentrations in fish from cold or warm waters he quickly established that salt provided only partial, if any, freeze protection. However, when he heated the blood of Antarctic fish to destroy all conventional proteins he found that the freezing point of the transparent fluid left behind was lower than that of the Southern Ocean. Clearly, something in that fluid was the magic ingredient DeVries had been looking for but, back in the 1960s, it

would take him another year to purify the concoction and to identify the molecules he found as glycoproteins, simple proteins wrapped up in sugars.

That alone was already a stunning achievement for a doctoral student, but he went on to describe the molecules completely, including their sequence of amino acid building blocks, using all the protein chemistry techniques available at the time. He named them, perhaps a little unpoetically, antifreeze glycoproteins or AFGPs.

By the time I first met DeVries, he had already spent 40 seasons on ice. Everybody at McMurdo Station and Scott Base, including people with absolutely no interest in fish, knew him and recommended that I talk to him. I found him in his office cubicle, poring over papers and notes, and not at all interested in chatting. It was only when I asked whether I could visit the aquarium that a reluctant smile crossed his weathered face.

DeVries helped design the facility to support his research, and he managed to keep several deep-water species, including toothfish, alive in tanks. He holds the record for hauling the first Antarctic toothfish out of the icy ocean. Earlier scientists had taken them from the mouths of seals and the stomachs of whales, but DeVries set up a winch system inside a small portable plywood shack and lowered the cable through the ice to the ocean floor. Over the years he has perfected his technique and has caught and released about 5500 of the giant fish as part of a tagging study.

The study provides much of the scant information about the species' growth, age and habits. Antarctic toothfish are the largest finfish in the Southern Ocean, but like most creatures that live in cold conditions, they grow very slowly, only about 5 centimetres (1.9 inches) a year. They can take eleven years to start reproducing. An average specimen weighs about 30 kilograms (66 pounds) but some older fish can be heavier and longer than the person trying to reel them through the hole in the ice. A 60-kilogram (130-pound) fish is probably more than 30 years old.

As we bent over a large tank at the McMurdo aquarium, a relatively small toothfish turned in slow circles, with its translucent fins moving like Japanese fans. Its bulging eyes, fat lips and strong jaws with inward-facing teeth made it look rather prehistoric, yet oddly fragile. The reduction of bone in favour of cartilage was clearly evident in the soft, bulbous skull.

Next to the tank, a dripping fish stomach hung from a clasp. It was part of an experiment to figure out if antifreeze proteins were present in all organs and where they might be produced. It also explained why everybody knew DeVries. Although he had examined almost all parts of the toothfish that had been killed in the name of science over the years, he had no use for the fillets. Invariably, the white boneless meat would find its way into the kitchens of McMurdo Station or Scott Base and turn the night's dinner into a gourmet feast.

A few years later, we met again, this time with his wife and colleague Chris Cheng. On this occasion, DeVries needed much less encouragement to talk and gave me an even broader smile when I asked him to recount his early days as a young scientist, countering the established view that salt had to play a major role in freeze protection. Even once he had discovered the antifreeze glycoproteins, some of his more eminent colleagues remained unconvinced. The only way to prove he was right was to show exactly how these molecules worked. He teamed up with Charles Knight, a cloud physicist from the National Center for Atmospheric Research in Boulder, Colorado. Knight knew all about hailstones, thunderstorms and ice crystals, DeVries knew a lot about Antarctic fish and their habitat. Together they figured out how to grow a single ice crystal in the presence of fish antifreeze. They found that the proteins attached themselves to the ice, much like antibodies bind to a pathogen or foreign substance, and divided the crystal surface into tiny areas that were so small and curved that the energy balance of the water molecules changed, stopping any other water molecules from binding to the ice crystal. The antifreeze works the same way inside a fish: the proteins wrap themselves around an ice crystal and stop it from growing to any size that could damage tissue. With the help of antifreeze proteins, fish can cool down to minus 2.6 degrees Celsius (27.3 °F), three tenths of a degree lower than the world's coldest ocean, without freezing.

A few years later, Cheng and her team found that Arctic fish produce a suite of similar proteins, which evolved independently in response to the same problem in a striking example of convergent evolution.

As significant as it was, the discovery of fish antifreeze proteins was only one piece of the jigsaw and prompted more questions than answers. Half a century after his first trip south, DeVries runs an ever-growing team of researchers, including Cheng, whose research with University of Auckland developmental biologist Clive Evans helped debunk another long-held view about the source of fish antifreeze.

Most biologists assumed that the antifreeze molecules were made by the liver, an organ known to produce a suite of other proteins. However, experimental results didn't quite add up. The team came up with a two-pronged approach, combining genetics with immunology, and developed specific genetic probes and antibodies to monitor the translation process that transforms antifreeze genes into antifreeze proteins. To their own surprise, they found no activity in the liver, but instead in the pancreas, oesophagus and upper stomach.

Evans, gregarious, chatty, and a veteran of two decades on ice, joined the DeVries team in 2000. In his own research he was using Antarctic fish as indicator

organisms to track the impact of pollution by measuring the activity of genes that are switched on in the presence of heavy metals. A similar approach, focused on the AFGP genes, helped solve the dilemma about where the proteins were made. The next obvious task was to work out what exactly happened once they had been activated.

Antarctic fish swallow ice with each gulp of water, and a few crystals enter the bloodstream through lesions in the skin or gills. Antifreeze proteins are produced in the exocrine pancreas, the larger of two parts that make up the organ, which also produces and secretes digestive enzymes. Evans and Cheng think that AFGPs are released into the gut like all other enzymes from the pancreas and protect the fish from ice crystals that come in with sea water and food.

A nearly identical mix of the proteins is also found in fish blood, but how the molecules manage to cross from the gut to the blood remains an open question. What is clear is that, somehow, protein-coated crystals end up in the spleen. In the gut, the antifreeze proteins bind to ingested ice to stop the crystals growing. Presumably, the ice is expelled in the faeces. The removal process is not that straightforward in the blood. Evans used nano-sized beads coated with fish antifreeze to mimic the journey of an ice crystal through the fish body. He ground up bits of every possible tissue in search of ice. The spleen is the only internal organ that returned positive results. The team's hypothesis is that the antifreeze proteins bind ice that gets into the blood but cannot destroy it. Instead, they prompt macrophages, cells that can engulf any foreign body, to swallow the entire protein-ice complex and move it out of harm's way within the spleen. There the ice accumulates until the fish finds a warmer spot in the ocean to melt its dangerous load.

Evans' investigations didn't stop with fish. He had noticed that many fish species carried leeches and so, out of pure scientific curiosity, his team put leech tissue through the same analysis. Sure enough, the leeches were full of fish antifreeze. Given the leeches' diet of fish blood it came as no surprise to find the antifreeze proteins in their gut. But it seems the parasites have managed to usurp their meal and distribute the proteins throughout their own body, including the sucker muscle, exploiting the freeze protection of their victims.

Slow motion

John Macdonald learned the hard way that Antarctic fish are as well adjusted to the chilling temperatures as is possible within the limits of biology. For more than four decades since the early 1960s he has returned to the ice so often he's lost count of the exact number of trips – or the days spent sitting in a converted shipping container, somewhere out on McMurdo Sound, fishing through a hole in the ice.

As an undergraduate student, he had seen some of the strange creatures biologists had fished up along the Antarctic coast during the International Geophysical Year, and so, like DeVries, he joined Donald Wohlschlag's group at Stanford University and headed south for a season. His research target was a fish called *Trematomus bernacchii*, known as bigheads to Shackleton's crew and as bernaks to many who study Antarctic fish. The small but big-headed pale brown fish are among the best-studied species, ubiquitous around the Antarctic coast and usually easy to catch in mesh traps almost identical to those used by Scott's men at the turn of the nineteenth century. Back in the 1960s, scientists baited traps with seal meat and measured the bernaks' oxygen consumption and metabolic rates in closed chambers at the McMurdo Station aquarium. Macdonald was one of the first to carry out his experiments in the field. He was particularly interested in how bernaks keep their brain and nervous system going in temperatures that would render any temperate fish comatose, if not dead. Borrowing methods used on Arctic animals such as the beaver, he carried out the first measurements on Antarctic fish, taken right there at the fishing hole to make sure they represented the natural environment as closely as possible.

Macdonald arrived in New Zealand in 1972, with a doctorate in neurophysiology and a young family in tow. Two years later, he returned to the ice to join DeVries' team for a summer and to continue his work on cold adaptation. By then, it was clear that the evolution of antifreeze proteins was an absolute requirement for survival in the Antarctic ocean, but Macdonald's focus was on a range of other adaptations that make life in the freezer more comfortable, at least for fish.

During his first few seasons, his fishing hut was close to Scott Base. Constant interference from other instruments, particularly blasts from the ionosonde (or chirpsounder, a special radar used to examine the ionosphere), made neurophysiological recordings difficult. When he towed his hut further out onto the sea ice, it became an impromptu laboratory, heated only by a small and temperamental diesel stove. But the warmer it was inside, the less happy the fish were, and so man and fish eventually reached a nippy compromise of 10 degrees Celsius (50 °F). Even in his retirement, Macdonald still remembers the patience and dedication it took to manipulate a tiny piece of tissue with nearly frozen fingers – and the laps he had to run around the hut, screaming, to work off pent-up energy, keep his body warm and his sanity intact.

Macdonald soon began collaborating with John Montgomery, then also a newly graduated fish biologist with an interest in the brain and sensory organs. Known as Mac and Monty, the pair set out to measure the speed of contractions in fish muscle fibres, the movement of eyes, and how fast Antarctic fish could swim if they had to. The results were clear: cold slows everything down. Physiologists working with cold-blooded animals have come up with the Q10 rule, which predicts that a change in temperature by ten degrees will result in a roughly twofold change in activity. Although Antarctic fish slow down, they defy that rule. They move at about

John Macdonald fishing at Terra Nova Bay in 2002. Macdonald's target was a ubiquitous fish known as bernak or *Trematomus bernacchii*. LEFT: COURTESY OF JOHN MACDONALD; RIGHT: ROD BUDD, NIWA

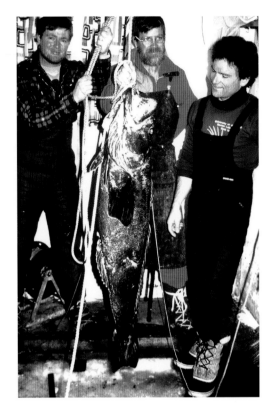

John Macdonald, centre, John Montgomery, right, and Gary Housley in the fish hut in 1985, with a toothfish. Macdonald retired from active research in 2006 to a farm north of Auckland but still visits Antarctica as a cruise-ship guide. Montgomery now heads the University of Auckland's Leigh Marine Laboratory.

One of John Montgomery's research discoveries was that Antarctic fish have a well-developed lateral line system, a sense organ that detects movement in the surrounding water, which allows them to capture prey in total darkness. For one species at least (borchs) this sensory system is tuned to the vibrations made by their planktonic prey. Working on the larval stages of silverfish, Montgomery also explored the relative importance of vision versus the lateral line sense. He found that vision dominates at first and that the eyes of a juvenile fish grow independently of the rest of the body. They grow rapidly during the first summer and continue to develop during winter, when the fish itself stops growing. The pattern stops after about three years, when the lateral line allows the fish to detect prey in the dark and continue feeding throughout the winter. Montgomery found that the lateral line system is sensitive enough to distinguish between different types of prey – acting like touch at a distance. COURTESY OF JOHN MACDONALD

half the speed of their remote relatives living in more temperate conditions, but that is significantly faster than those warmth-loving fish would if they happened to be tossed into icy water.

Life in the cold turns out to be a trade-off between biology operating as quickly as possible yet not so fast that systems start collapsing. It is all about flexibility and fluidity, all the way down to each cell and molecule. Each of the millions of cells that make up a fish, or human for that matter, is surrounded by a double-layered membrane of fatty acids. Its fluidity depends on the ratio between saturated and unsaturated fatty acids – much like the difference between butter and oil. The lipids of Antarctic fish cell membranes include far more unsaturated and shorter fatty acids, making the cell structure more elastic and supple. The molecular bonds between the membranes and any enzymes that act as channels into the cell's interior are just loose enough to operate efficiently without breaking. In nerve cells this means that the electric impulses between the brain and muscle or sensory organ travel faster; in other cells it means that any chemical transport across the membrane works better.

However well adapted they may be, Antarctic fish pay a high price. As DeVries found early in his work when he transferred his catches into slightly warmer water, the bodily chemistry of many Antarctic fish species is so well tuned to operating in sub-zero temperatures that most don't survive beyond 6 degrees (43 °F) plus.

Evolution is often seen as a process that moves forward by incremental improvements, but Antarctic fish survive on a knife edge between nearly fatal but ultimately useful changes and secondary adaptations that compensate for those changes. Montgomery has devoted some of his research to exploring how a disadvantage can be the beginning of a new evolutionary branch, finding that icefish in particular illustrate this precarious lifestyle. The loss of haemoglobin would have been fatal under any other circumstances except in extremely cold water. He argues that far from being an advantage, it was merely a tolerable change. The fish survived because oxygen is highly soluble in cold water but they had to supersize their hearts, widen their vessels and produce more of their colourless blood to make up for the drastic change. From a functional point of view, icefish took a step backwards, but they were able to flourish partly because of the lack of competition in their new niche.

Running on fat

The view from the Scott Base mess takes in the Ross Ice Shelf on one side and a vast sheet of sea ice on the other, with Black and White islands and Minna Bluff as permanent signposts to the Transantarctic Mountains in the background. The seasonal sea ice is 2 or 3 metres (6 to 10 feet) thick, and where it meets the land it buckles and pushes up in dramatic pressure ridges. Huddled against this striking backdrop of sculpted ice is a small hut, painted in the trademark Scott Base green. The 'wet lab', a step up in comfort from the makeshift fishing shelters out on the sea ice, houses a small sample of vertebrate life fished up from the ocean below.

During my Antarctic visit in 2006, this was where I learned my first lessons in Antarctic fish taxonomy. Pipes gurgled as ocean water flowed through each tank, and pink and brown fish nestled in crevices or perched on top of rocks, fins splayed. I had come looking for Victoria Metcalf, a young biochemist studying fat metabolism in fish, and I found her at the back of the hut, bent over a laboratory bench, busy sorting through samples of frozen tissue and blood. As she showed me around her collection of aquaria, I was introduced to fish from several taxonomic suborders with different cold adaptation strategies. Bernaks, included to represent the dominant notothenioid group which produces antifreeze glycoproteins, were the

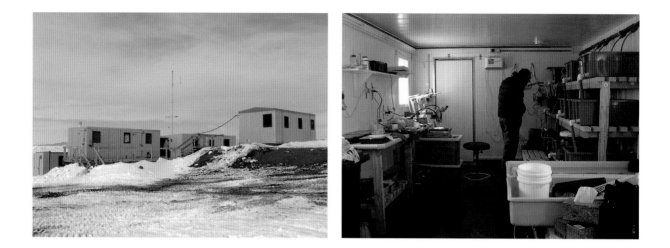

The wet lab at Scott Base with its tanks enables scientists to work more comfortably than in a fishing hut. ANTARCTICA NZ PICTORIAL COLLECTION: BILL DAVISON/K057 07/08

most familiar. Eelpouts, some of the most abundant non-notothenioid fish in the Antarctic ocean, were of interest to Metcalf because they use a different set of chemicals as freeze protection. A snailfish provided an example of a group of fish that go without. Perhaps it was also a nod to DeVries, one of Metcalf's mentors. Snailfish belong to a little-known genus of small, jelly-like, deep-ocean bottom-dwellers which simply increase the sugar load in their bodily fluids to survive by supercooling. Their strategy is risky. They would snap freeze on contact with even the smallest ice crystal but their habitat at depths of more than 500 metres (1600 feet) is usually free of ice because pressure lowers the freezing point of the water. One such pink scaleless fish was caught in McMurdo Sound sometime during the late 1970s but nobody took the trouble of identifying it. DeVries sent it to a Russian taxonomist, who confirmed that it was new to science and named it *Paraliparis devriesi,* following the old taxonomic tradition of naming new species after eminent researchers or explorers. DeVries is amused by the irony that the discoverer of fish antifreeze should lend his name to a fish that doesn't need it.

Metcalf came to study Antarctic fish by serendipity. Her initial interest was in the evolution of albumin, a ubiquitous protein most vertebrates, including humans, use to transport fat around the body. She transferred her attention from other animals to Antarctic fish when a colleague handed her samples of toothfish blood and she found no trace of albumin. Like haemoglobin in icefish, this was another common protein that seemed to be of no use to fish in icy waters. However, in this case, Antarctic fish are not alone. Albumin is also missing from the blood of several more temperate species, including New Zealand thornfish and spotties, suggesting a more ancient evolutionary step.

The discovery left Metcalf with the puzzle of how Antarctic fish deal with fat transport. Lipids are central to their life. Apart from providing buoyancy, they are the primary fuel, as these fish prefer fat to the carbohydrates most other vertebrates run on. The question on Metcalf's mind was whether this was somehow tied up with their need to protect themselves from freezing. Toothfish use lipoproteins to ferry lipids around their body and Metcalf speculates that other polar fish species may have replaced albumin, which is found in high concentrations in other vertebrates, with molecules that work at lower concentration to keep their blood thin enough to flow.

Metcalf has also expanded her research to investigate another aspect of lipid metabolism. She was curious to find out if Antarctic fish could modulate the ratio of unsaturated to saturated lipids to change the fluidity of their cell membranes in response to slow temperature changes. The Antarctic ocean may be extremely cold, but it is also one of the most thermally stable environments. Temperature changes are minimal throughout the year, and evolutionary theory suggests that, by adapting to life in the freezer so well, Antarctic fish may have lost their ability to cope with variation. There is one well-studied exception, however. Given enough time, *Pagothenia borchgrevinki*, a small spotted fish named in honour of the Norwegian explorer Carsten Borchgrevink, can acclimatise to ocean temperatures of up to 4 degrees Celsius (39 °F). Known as borchs to biologists, these fish are among the few that have left the seafloor and taken up swimming again, feeding on krill and other small crustaceans just below the ice. They are far more active than the sedentary bottom-dwellers and, so the hypothesis goes, may therefore also be more tolerant of higher temperatures. In contrast, bernaks show no capacity for acclimation which, in the light of anthropogenic climate change, is making Metcalf and other fish biologists worry about the future of Antarctica's unique fish fauna.

Plunging into the unknown

The prospect of diving under Antarctica's sea ice is both daunting and exhilarating, even for the most experienced divers. The intense concentration was palpable as I watched ecologist Simon Thrush and dive supervisor Rod Budd squeeze into their fortified dry suits, with a double-lined hood that sealed tightly to the mask. Usually convivial, both men had gone quiet, except for the necessary checks with their support team. As movement was becoming increasingly cumbersome, others had to help them pull on a second pair of gloves and slip on an extra tank and regulator for safety, in case any of the valves were to freeze open in the frigid waters below.

The narrow entrance hole leads through several metres of ice to the ocean below. ROD BUDD, NIWA

They slid down the edge of the 1-metre (3-foot, 1-yard) entrance hole in the ice, hovered briefly to check for any leaks in their suits or equipment, and slowly disappeared into an alien world. Their only connection and means of communication was a 50-metre (160-foot, 55-yard) lifeline. The dialogue was simple. A long pull on the rope was answered in kind from the surface and meant that everything was okay. Four short pulls, known as the 'bells', requested more or less slack on the rope, but any more than four quick tugs and the support team knew the diver was in trouble and had to be pulled out as quickly and as safely as possible.

It was early December in 2001, and this was one of the team's first diving expeditions to explore Antarctica's underwater world. They were not looking for fish. Their focus was on the improbable spineless creatures that somehow manage to thrive in the merciless conditions underneath the sea ice and their interactions with their environment. This southern summer marked the start of the long-term IceCUBE project, a comprehensive study of life on the ocean floor, which

ABOVE LEFT Final checks before a dive. ROD BUDD, NIWA

ABOVE RIGHT Two divers are connected via lifelines to their support crew on the surface, while a back-up diver waits, kitted out, ready to jump in if emergency strikes. PETER MARRIOTT, NIWA

would eventually visit several other sites along the coastline to observe benthic, or bottom-dwelling, communities along the latitudinal gradient. The team had set up a tent camp at Cape Evans on Ross Island and dragged a converted shipping container out onto the ice to cover the dive hole. A few yards further, several other holes provided alternative access or escape routes in case a Weddell seal decided to use one as a breathing haul-out.

Everybody on the surface remained on tenterhooks throughout the 30-minute dive, but the divers relaxed as soon as they had passed through the narrow channel of ice and their eyes had adjusted to the dimmed light and exceptionally clear water underneath. For most of the year, underwater visibility around Antarctica can reach 200 metres (660 feet, 220 yards) and it can be disorienting and difficult to judge size and distance. But the divers had a clear target. During an earlier descent, the team had already pegged out a white line along the rocky volcanic seafloor, and the job now was to move along that transect to collect video footage and sediment samples to get a better idea of the diversity and abundance of invertebrate life on and within the seabed.

Some days later, back at Scott Base, Simon Thrush was still buoyed by his time under the ice as he showed me the video footage of masses of bright pink starfish, giant worms and purple urchins. In contrast to the near monochromatic images of the land, the seafloor was teeming with colour and life.

Thrush's research focus is on coastal habitats, and for him, the ocean floor off the Antarctic coast is one of a very few places on Earth that remains largely unmodified by people. When he talks about it, he brims with the joy of knowing that he is one of a privileged few who have entered this pristine world.

Suspended between the algae-coated ice and the ocean floor. ABOVE: PETER MARRIOTT; OPPOSITE: ROD BUDD, NIWA

The earliest descriptions of Antarctic seafloor communities go back to Scott's first collector, Thomas Hodgson, who gathered specimens from depths of more than 200 metres (660 feet) from aboard the HMS *Discovery* in 1901. A year later, Willy Heinrich, a carpenter with Erich von Drygalski's German expedition on board the *Gauss*, became the first person to dive under Antarctic ice, in the middle of winter and with surface temperatures of minus 30 degrees Celsius (-22 °F).

More than half a century later, US Navy divers first used scuba equipment during dives to repair ships and test equipment, but the first person to dive under ice in the name of science was a young laboratory manager called Verne Peckham, who began collecting seafloor animals in 1961. He made 35 dives, down to 46 metres (150 feet), during the dark winter months, by himself and for no reason other than curiosity. Art DeVries helped him cut a hole in the ice with a chainsaw.

Two years later, Paul Dayton, now a professor at the Scripps Institution of Oceanography, began his pioneering scientific diving trips under ice, logging hundreds of dives and descending to depths of up to 60 metres (200 feet), beyond what is now considered safe. His work laid the foundation for marine ecological research along Antarctica's coastline and sketched the first maps of changing underwater habitats in McMurdo Sound and along the Ross Island coast. Despite the colourful display, he found that very shallow sediments a few yards below the sea ice are often relatively barren cobblefields compared to the denser communities of starfish, sea anemones, sea urchins and soft corals that cover the ground from about 15 metres (50 feet) of depth. Massive colourful sponge gardens, with some individuals the size of a car, begin to dominate below 30 metres (100 feet) until they give way to large aggregations of bryozoans, tiny animals that build up complex colonies similar to corals in tropical waters.

Dayton focused his diving excursions on two very different areas. Along some stretches of the Ross Island coastline, on the eastern side of McMurdo Sound where nutrient-laden currents sweep down from the north, he found that each square metre of sediment supported more than 150,000 organisms. In contrast, on the opposite side of the sound, water emerges from below the Ross Ice Shelf with little in the way of food and life is more akin to that in the deep ocean.

Simon Thrush credits Dayton with inspiring his interest in Antarctica's marine ecology. When he returned to New Zealand in 1985, Thrush had already spent a total of three months under water as part of his doctoral project on the seafloor ecology off the Irish coast. He was working at the marine laboratory at Portobello, a small field station of the University of Otago, when Dayton showed up unexpectedly. Freshly back from a summer season in Antarctica, he was visiting Dunedin in pursuit of his other research interest in kelp forests.

The impromptu meeting sparked a lifelong collaboration. Thrush soon joined Dayton's Antarctic dive team for two seasons and returned enthused with the idea of setting up a similar research project in New Zealand, focusing on coastal ecosystems and biodiversity in Antarctica.

Many years later, at Cape Evans, Thrush's enthusiasm for Antarctica's underwater ecology was undiminished as he showed me more footage of a seafloor swarming with life. To my uninitiated eye, the creatures crawling across the video screen looked similar to those you would find along more temperate coastlines elsewhere, but Thrush was quick to point out that the composition of the communities is distinct and unique to Antarctica. Some major groups of animals, including crabs, are missing completely, while others flourish in numbers not seen in warmer waters. The proportions shift in favour of large sponges and soft corals,

Taking samples on the ocean floor. ROD BUDD, NIWA

which provide additional habitat and shelter for the more complex creatures that graze on them. Hundreds of starfish carpet the seafloor, mingling with sea urchins and giant carnivorous nemertean or ribbon worms that grow to more than a metre (3 feet, 1 yard). In the sub-zero (sub-32 °F) temperatures, all creatures grow slowly, but that means that they can grow very old and impressively large.

Some of their adaptations are equally unique to polar waters and mostly developed to cope with the extreme fluctuations in food supply. As in fish, the low temperature slows down the metabolic rates of invertebrates and they essentially tick over living on next to nothing and feast whenever food is plentiful. But some species have evolved tricks to cope better during the famine. Sponges are among the most primitive sea animals. They have no nervous or digestive system and

The ocean floor off the Antarctic coast teems with life and colour. **CLOCKWISE FROM TOP LEFT** Urchins are often covered in debris as a form of camouflage and added protection; a feather duster worm, *Potamilla antarctica*, and starfish, *Odontaster validus*, at Cape Evans; carnivorous nemertean worms – they can grow up to a metre (3 feet); and an *Odontaster validus* starfish. A scavenger, it will also attack much larger starfish by climbing on top and everting its stomach. Once the acidic digestive juices have burnt a hole into the prey, other members of the same species sense that there is food and join the fray. ROD BUDD, NIWA feed simply by filtering the currents that pass by. More than 300 different species have been described in the Ross Sea, and most of them are glass sponges, normally found in the deep sea, that weave their skeleton from silicon and are covered by a scaffold of tiny sharp points called spicules. When food becomes scarce, these simple suspension feeders will take in algae to boost their energy requirements, but one particular species has taken the symbiotic tenancy arrangement one step further. It uses its glass spicules as natural fibre-optic cables, to conduct light to its photosynthesising guests.

Sponges remain a significant component of the seafloor even after death. Their spicules form a complex mat, increasing the surface area and providing habitat for smaller plants and animals. Larger, more complex animals have also adapted to the feast-famine cycles. When their usual food becomes scarce, small, usually herbivorous, starfish will gang up on a much larger sponge-eating starfish and pour digestive juices on its skin. The tiny leaking wounds attract other starfish and ribbon worms and, collectively, these unforeseen predators overwhelm the much larger prey.

Frozen in motion

Ice determines life under water. Algae and larger seaweeds need light to grow and the amount of sunshine that filters through the ice into the water column during the short summer is a key factor in their survival. These tiny marine plants are the primary producers, using photosynthesis to turn carbon that is dissolved in the ocean into larger molecules. All organisms that follow along in the food chain depend on them.

At Cape Evans, brown and green smudges on the underside of the ice signify flourishing populations of single-celled algae, but light penetrates even further, cultivating patches of algae that encrust the rocks and sediments on the seafloor, and even two kinds of seaweed with leafy fronds. These seaweeds survive just a little north of their extreme southern limit at Hut Point, also on Ross Island, near McMurdo Station.

From a plant's perspective, life in Antarctica is a long day followed by a long night. When the sun dips below the horizon around the end of April, the seaweeds survive the perpetual darkness of winter by shutting down growth and persisting on energy stores built up during their short bursts of photosynthesis when light is available. Sea-ice algae die off and drop to the seafloor, turning into an important food source for diverse bottom-dwelling communities.

The amount of sunshine and sea-ice cover are two factors that vary predictably with latitude, and the Cape Evans team expanded its research to explore several other sites along the coast of Victoria Land, which spans from 72 degrees South at Cape Adare to 86 degrees South at the southern tip of the Ross Ice Shelf. Led by benthic ecologist Vonda Cummings, the group's spotlight remained on life forms on the seafloor and their ecology, but to engage in a broader-scale analysis, the team also joined efforts with other research groups to form part of the Latitudinal Gradient Project. This international collaboration brings together people studying terrestrial, marine and freshwater ecosystems, as well as the physics of sea ice and glaciers, in an endeavour to understand how habitats and environmental conditions change the deeper south you go.

Some years after visiting Cape Evans, I met Cummings in her office at the National Institute of Water and Atmospheric Research (NIWA) in Wellington to see if any patterns were beginning to emerge. She picked up a scallop shell from a cardboard box and flicked to an image on her computer screen of densely packed scallops on a rocky ledge. Shallow beaches are where you would normally expect to find scallops, but at Terra Nova Bay the team had found them clinging to rocks, holding on with the sticky anchoring threads known from other shellfish, but not from scallop species that live along New Zealand's coastline. When Cummings brought some Antarctic scallops back to a tank in her laboratory, they preferred

propping themselves up on each other or other objects rather than lying down flat like their temperate relatives.

The scallop shell she handed me was noticeably thin and brittle. This is typical of all molluscs that live in polar regions, because colder waters hold less calcium carbonate, an essential building block for shells, spines and other skeletal structures. With just one of the many species that make their living on the ocean floor around Antarctica, Cummings was demonstrating both the robustness and fragility of the ecosystem.

Cummings is the only IceCUBE team member not qualified to dive, but after analysing hundreds of hours of video footage, handling hundreds of specimens and hanging on to safety lines during many underwater explorations, she feels she has had equal visiting time under the ice. The team has since explored eight further sites along the coast, collecting video footage and sediment samples and trying to match the abundance of invertebrate fauna on the seafloor with that of smaller organisms found living in the sand. They have also analysed the sediments, algae, and tissues of various species for stable isotopes of nitrogen and carbon to figure out who eats who.

It came as no surprise to the ecologists that latitude per se is not a useful measure for predicting variation in the seafloor community. Life at the bottom is influenced most strongly by sea-ice conditions, which directly govern primary production

Scallops use each other as substrate. In the photograph on the right they are gathered in a moat near a standing wave of ice.
ROD BUDD, NIWA

Each of these Plexiglass chambers seals a brittle star off from its environment. They are hand-fed algae by the science team in an experiment to work out the relationship between food producers and consumers in Antarctic waters.
ROD BUDD, NIWA

and the amount of food available by limiting the availability of light. Nitrogen isotope signatures, used to determine a creature's level in the food chain, showed that some bivalves and echinoderms, including sea urchins, can shift their diet when needed. They grazed on freshly produced algae in ice-free areas, but then switched to dining on detritus in sites further south that are covered by more permanent sea ice. Clearly, they were able to adjust their food-chain status depending on what was on the menu.

The thickness and duration of sea ice proved to be the main drivers, while the physical structure of the seafloor habitat, ranging from boulders to rocks to fine sand, played a lesser role. Combined with snow cover and the amount and quality of food available, these factors explained almost all of the variation in the composition of the seafloor communities between different habitats. However, the interactions between them are far from simple. Even a subtle change somewhere in the environment can trigger a cascade of other changes with ramifications for the entire ecosystem.

Ocean diving, Antarctic style. ROD BUDD, NIWA

Predicting how the entire ecosystem might react to even one such change is a challenge, but for Cummings and Thrush, the ultimate goal is to understand life on the seafloor well enough to foresee the impact of changing climate and ocean conditions, and to offer advice on how to maintain the resilience of marine ecosystems. Thrush expects that warmer ocean temperatures might open up new food sources as the sea ice recedes, but at the same time could make life harder for sponges and other simple filter feeders that act as nurseries and sanctuaries for many of the more highly developed species. If sponges were to diminish or even disappear, these larger animals could be forced to follow suit.

An equally important concern for the team is the acidification of the oceans as they take up more carbon dioxide from the atmosphere. Although the sources of carbon dioxide are elsewhere, the acidifying effect is strongest in polar regions, because it dissolves more easily in colder water. In a more acidic ocean, and one where levels of calcium carbonate in the water are already relatively low, scallops and any other skeleton-building seafloor-dwellers would struggle even more to build their shells, and the already brittle structures would be more likely to dissolve. The changing ocean conditions could also open the door to invasions by new predators, including crabs, which would find the brittle shells of many bottom-dwellers easy barriers to crack.

Inventory of life

On the last day of January 2008, NIWA's research ship *Tangaroa* headed south for a 50-day voyage bound for the Ross Sea. On board was an international team of marine biologists and technical staff, an experienced crew captained by a veteran of several expeditions into the wild Southern Ocean, various sizes of trawl nets and sampling devices, and an especially designed contraption festooned with high-definition video and still cameras.

The ship encountered the worst sea-ice conditions in 30 years. It took only two days for sheets of ice to form across the entire eastern Ross Sea. The *Tangaroa*'s deck and equipment froze over completely and, with temperatures hovering around minus 20 degrees Celsius (-4 °F) and the air saturated with tiny ice crystals, it became almost impossible for the scientists and crew to continue working.

NIWA's deep-water research vessel *Tangaroa* working in heavy pack ice in the Ross Sea.
PETER MARRIOTT, NZ IPY-CAML

Nevertheless, they returned with a rich bounty, including samples from several sites that had never been explored before. The nets had been designed to sample marine life of all sizes and hauled up everything from the tiniest phytoplankton to giant toothfish. Almost every catch included something new. All in all, the team returned with more than 37,000 specimens. The invertebrates alone covered 51 groups; the total fish catch included possibly nine new species.

Tangaroa's voyage was part of the international Census of Antarctic Marine Life (CAML) – a massive exercise to draw up an inventory of all residents in Antarctic waters – which in turn is part of an even larger effort to catalogue life in all oceans. Initial findings published in 2010 confirmed that Antarctica's coastline and oceans are home to many species that are found nowhere else. In the case of gastropods, for example, three out of four species are endemic to the region.

The *Tangaroa* encountering pack ice in the Ross Sea in 2008. The extent of sea ice was greater than had been seen in 30 years, but the crew and science team returned with a rich bounty, including more than 100 new species. Long voyages such as this one often inspire scientists and crew into creativity – including poetry, art and photography. GLEN WALKER, NZ IPY-CAML

NIWA fisheries scientist Brent Wood (left) and Italian Antarctic Museum taxonomist Stefano Schiaparelli (right) examining an Antarctic toothfish in *Tangaroa*'s wet lab. Antarctic toothfish, *Dissostichus mawsoni*, are named after Australian explorer Douglas Mawson who sailed with Ernest Shackleton on the *Nimrod* and was involved in the first ascent of Mount Erebus. *Trematomus bernacchii* is named after Louis Bernacchi, who was part of Scott's 1901 foray to Ross Island. And *Pagothenia borchgrevinki* honours Norwegian Carsten Borchgrevink, the commander of the British Southern Cross Antarctic Expedition of 1898, which established the first winter station on the continent. RICHARD O'DRISCOLL, NZ IPY-CAML

Almost 10,000 species have been identified in the Ross Sea but scientists expect to find many more.

The Ross Sea represents only a small part of the Southern Ocean but it is one of its most productive and largely intact areas. It is as far south as one can go by sea. The Ross Ice Shelf, a floating extension of the glaciers that flow down from the Polar Plateau, covers its southern end. Beyond this icy blanket, the Ross Sea extends across Antarctica's continental shelf and the slope, where the ocean floor drops suddenly down to 3 kilometres (9800 feet) in depth. Expeditions stretch back to the voyages of James Clark Ross, the explorer who discovered and named not only the ocean and giant ice shelf, but most other features along the coastline.

Strong currents upwelling from the deep ocean bring plenty of nutrients to the shelf, and both the water column and seafloor are teeming with life. At one end of the food chain, phytoplankton flourishes, while at the other end, top predators

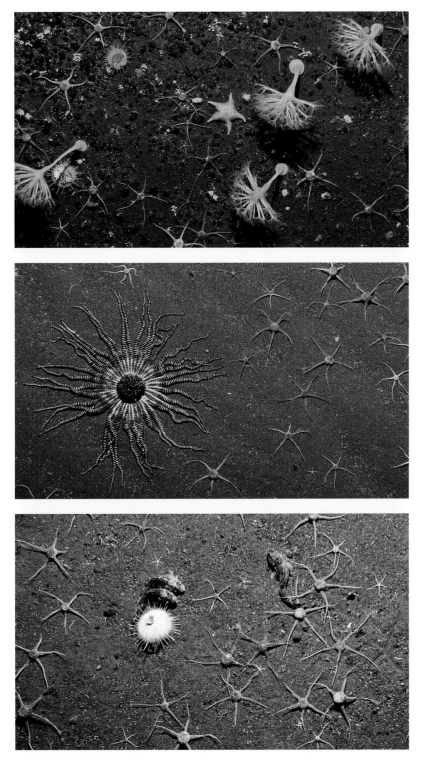

Life at 150 metres (490 feet) deep, photographed near Scott Island, a seamount in the northern Ross Sea: abundant sea pens and brittle stars; a starfish (*Labediaster* species) surrounded by brittle stars; and seafloor-skimming notothenioid fish. DTIS CAMERA, NZ IPY-CAML are in abundance. Three million Adélie penguins – more than a third of the entire population – feed and breed in the Ross Sea region. It is also home to almost a third of the world's Antarctic petrels and emperor penguins, half the entire Weddell seal population, and substantial numbers of Antarctic minke whales, leopard seals and orcas, including a population of the latter specially adapted to feed on Antarctic toothfish. Some scientists have come to think of the Ross Sea as the Serengeti of the oceans and they want to see it fully protected from any commercial exploitation, including fisheries. At the helm of the protection campaign are two penguin ecologists, New Zealander Peter Wilson and American David Ainley.

They go back a long way. Between them, the pair counts more than 70 summers on ice, focused on taking stock of the fluctuations in Adélie populations, their breeding success and survival rates, and trying to understand the birds' ecology. Ainley is quiet, almost withdrawn. Wilson smiles and talks easily. Despite their opposing personalities, each is passionate about the smelly and aggressive penguins and the Ross Sea, and their importance as an indicator species and a priceless natural laboratory respectively. Even whalers and sealers stopped just short of the southern parts of the Ross Sea. Since then, commercial interest has remained limited by its inaccessibility; until 1996, when the first fishing fleet arrived, led by New Zealand, to catch Antarctic toothfish.

For Ainley and Wilson, the Ross Sea represents one of a very few places on Earth where scientists have managed to gather information about the environment and its inhabitants pre-dating human impact. Almost anywhere else, ecological studies are compromised from the start by this lack of baseline data, but the Ross Sea has yielded enough insights for scientists to draw a compelling picture of its natural state.

Their collaboration began with the first aerial surveys of Adélie colonies during the summer of 1981/82. That season marked the start of an international survey of Antarctic seabirds around the entire Antarctic coast. Wilson had already flown small aircraft in the name of ornithology, monitoring gannet colonies in the North Island of New Zealand and tracking kaka in the South Island. He reckoned aerial surveys of Adélie penguins, carried out at the right time of the breeding season when birds were sitting on nests, would deliver the most reliable headcount and a good estimate of the size of each colony.

By the time Wilson arrived at Scott Base to launch the first flights, two determined young ecologists had already started ground-based observations of Adélie penguins more than twenty years earlier. In November 1959, Rowley Taylor and Euan Young jumped out of a helicopter at Cape Royds, a rocky promontory on the western flank of Ross Island and the penguins' southernmost breeding colony.

Cape Royds is arguably the most visited colony, not just because of the penguins themselves but because it is where Ernest Shackleton built a hut for his 1907–09 Nimrod expedition. He had already been to Antarctica a few years earlier as a member of Scott's Discovery party but was sent home early with scurvy. This was the first of three Antarctic expeditions he led himself. It took him almost, but not quite, to the South Pole. Famously, Shackleton turned his men back 156 kilometres (97 miles) from their destination, at 88 degrees South, when bad weather and dwindling supplies threatened their safe return. They had nevertheless ventured further south than anybody else at the time. When they returned to Ross Island, they departed from Hut Point, without going back to Cape Royds, but Shackleton had left instructions with the rest of his shore party to leave a note for any future expeditions. The letter listed all provisions and equipment, enough to last fifteen men for one year, and details for the coal store. It ended with an invitation to any subsequent party to use whatever supplies they needed.

Spending the southern summer studying Adélie penguins and polar skuas and the relationship between the two remarkable seabirds, Taylor and Young took

Cape Royds

A skua flies off with an
Adélie egg. ANTARCTICA
NZ PICTORIAL COLLECTION:
J WAAS/K027 04/05

Shackleton at his word. They cleared the snow and ice from the hut, cranked up the
stove with a coal fire, and made themselves at home. From their vantage points,
they observed as both birds went about their lives and tried to raise their broods.

Adélies and skuas are Antarctica's most easily accessible birds, with a long history of observations stretching back to the very first scientific expeditions. Both
need ice-free patches of land on which to breed and often establish overlapping
colonies, with the penguins nesting close together while the skuas are more widely
dispersed. Like smaller gulls elsewhere, skuas are predators and scavengers and
are often seen patrolling the penguin colonies in search of eggs and dead or live
chicks. Their own eggs are laid weeks after the penguins begin incubation; and
by the time skua chicks hatch, penguin chicks are already the size of a good meal.
For a long time biologists thought that the relationship was clear: skuas eat young
Adélies. Images of hapless penguin chicks being clobbered to death by the clumsy,
screeching gulls, while adult penguins stand by flapping their sturdy wings use-
lessly, have only reinforced the skua's reputation as the villain of the Southern
Ocean.

Young's observations rehabilitated the skua, at least partly. Despite their bad
image, skuas turned out to be rather incompetent predators of penguins and
their consumption of eggs and chicks was mostly the result of scavenging. Only
skuas nesting close to or within penguin colonies fed on their neighbours. The rest

Adélies treasure pebbles, the only material available to build their nests.
ANTARCTICA NZ PICTORIAL COLLECTION: RACHEL BROWN; GUS McALLISTER/K002 03/04

foraged at sea, dive-bombing for small Antarctic silverfish and displaying rather more skill and elegance in flight than on land. Skuas also showed capacity for learning, becoming almost tame as they got used to the regular human visits. In contrast, Adélie penguins seemed to have only one mode of action: vicious attack, no matter whether the intruder was human or avian. In terms of aggression in defence of offspring or territory, the uneasy neighbours were on a par.

Named after the wife of the nineteenth-century French explorer Jules Sebastien Dumont d'Urville, Adélies are extremely curious and fearless despite their small size. They build their nests close together, just outside pecking reach, and engage in almost constant territorial squabbles with their neighbours. The colonies are always in motion. Even when the birds are incubating eggs, they rarely sit completely still or stop calling, sky-pointing their beaks and trumpeting a throaty string of shrieks. Vocal recognition is an important part of Adélie communication, but so is posturing. A straight or one-sided stare, amplified by bright white eye rings, displays aggression. Crouching, with the eyes held down to expose more white, delivers an even clearer message of dominance.

Pebbles are treasured possessions. Adélies build their nests out of carefully selected and placed stones, and they guard them fiercely. Sometimes, they return from foraging trips at sea with a pebble gift for their mate, and they will pilfer stones from neighbouring nests, sparking yet more boundary disputes.

By the time of my first visit in November 2001, the code of conduct near historic huts had changed dramatically. Only a handful of specially trained guides were authorised to take people into any of the remaining Heroic Era huts on Ross Island, including Shackleton's Nimrod hut and Scott's huts at Cape Evans and Hut Point. Backpacks were off limits inside. Everybody had to scrub their boots outside to avoid bringing in any ice or scoria and, inside, nothing could be touched. The historic huts had survived a century on ice but snow drift and melt water were beginning to cause damage and their conservation had become a priority. Despite the restrictions, the experience was spine-tingling.

Shackleton's hut is light and open, with the stove as its centrepiece still exuding a sense of warmth even decades after the last fire has gone out. Unlike Scott's expeditions, Shackleton's was not a naval venture and there is no separation or ranking evident in the hut's layout. Long woollen socks are stretched across some bunks, and the galley shelves are packed with tins and cans of preserved foods.

When I returned five years later, conservators had removed snow and ice from underneath the building and waterproof cladding around the perimeter had stopped melt water from entering. Temporary windows and doors that had been put in place during the 1970s had been replaced with historically more accurate materials. It looked even more inviting, alive. Leaking cans of food that had been removed for preservation had yielded century-old dehydrated parsnip, red cabbage and rhubarb still intact, and caramelised citrus fruit still redolent of lemon and lime. No doubt, Shackleton had chosen Cape Royds so that his men could supplement their canned provisions with fresh penguin meat and eggs as well as for the morale-boosting entertainment of living next door to a noisy penguin metropolis.

A penguin summer

Although gentoo and chinstrap penguins breed along the coast of the Antarctic Peninsula, Adélies and emperors are the only true Antarctic species. They rely on sea ice for their survival. Emperors breed on the frozen ocean, and both species need ice floes as foraging platforms. Of the two species, Adélies are clearly easier to observe. Unlike the winter-breeding emperors, Adélies arrive at their colonies in late October, ready to establish nests and to raise their chicks before the end of the short Antarctic summer, in synchrony with the freezing and thawing cycle of the sea ice that surrounds the continent. By early December, all eggs have been laid and the females have returned to sea to feed. The only penguins on land are males sitting on nests – ideal targets for obtaining an accurate count from the air.

OPPOSITE The stove is the centrepiece of Shackleton's hut at Cape Royds.

A skua pestering an Adélie penguin at Cape Bird.
ROB McPHAIL

Rowley Taylor returned to Antarctica with Peter Wilson to launch the annual fly-overs in 1980. Initially the focus was on colonies on Ross Island, which could be monitored by helicopter each breeding season. These annual flights have continued ever since, building on the counts previously collected on the ground; and the data set charting the penguins' population size and breeding effort since 1959 is now the longest-running for any Antarctic creature.

The Ross Island surveys soon expanded to include regular flights, every three to four years, along the entire Victoria Land coastline, all the way north to Cape Adare and the largest Adélie colony with more than half a million birds. During the early years, the only way Wilson could get a plane to fly that route was to hijack a fully loaded C-130 Hercules at the start of its return journey from Antarctica. It didn't make his team popular. With cargo stacked high in the back of the plane and passengers kitted out in Antarctic survival gear strapped into narrow folding seats along the sides and the front, this was already a long and uncomfortable journey. Whenever the penguin team came on board, it turned into a twelve-hour mission of aeronautical gyrations. Wilson was often the only person who didn't mind. Kneeling on the floor next to the open door, with air screaming past, he would fiddle to reload rolls of film into a large-format camera while directing the pilot through 180-degree turns. With G-forces remodelling his face, he leaned out to take a series

A skua with its chick at Cape Bird (above left). Adélie chicks hatch a few weeks before skua chicks.
ROB McPHAIL

of images of each colony. The effort and lessening of regard was worth it. During the first flights, the team discovered eight new colonies – and they soon convinced the Royal New Zealand Air Force to provide flights dedicated solely to research.

The annual helicopter survey on Ross Island focuses on three colonies: the smallest, with only about 2000 breeding pairs, at Cape Royds, and two larger ones at Cape Bird (60,000 breeding pairs) and Cape Crozier (150,000 pairs). Each season, biologists also hunker down in small huts to continue ground-based observations at each site. While David Ainley's team sets up at Cape Royds and Cape Crozier, Cape Bird is the summer residence for New Zealand penguin researchers. Kerry Barton has spent countless summers there, dwelling in a small shelter perched on a slope above a long stretch of volcanic beach. The ice-free shore offers ideal territory for penguins and biologists alike. Framed by the massive Bird Glacier to the north and vertical cliffs to the south, it is packed with more than 100,000 birds at the height of the breeding season. For Barton, it offers a smelly but welcome change of scene from the weeks spent focused on the painstaking task of analysing the aerial images for tiny black dots. Here, she can observe first-hand what makes the difference between a good or bad breeding season.

A small area of the northern colony is fenced off and the penguins whose nests are inside the enclosure carry microchips under their skin. As they scuttle between

The beach at Cape Bird is
home to 60,000 pairs of
Adélie penguins and their
chicks. ROB McPHAIL

their brood and favourite ocean launching pad, they have to cross a weighbridge each time they enter or leave the area. A data logger records their ID and weight, allowing researchers to monitor how much food they were able to bring in for the chicks, how long they have had to forage for it, and the frequency of their trips. Other birds carry small satellite tags that monitor where and how far they are travelling. Time-depth recorders keep track of when they are diving, how deep they go, and how long they stay under water. All this information combined produces a comprehensive assessment of how much energy it takes a pair to raise their offspring.

Adélie penguins nesting at Cape Royds, the southernmost colony, often face harsh seasons when the sea ice doesn't break up in the southern McMurdo Sound. In those years, foraging birds have to feed through cracks in the sea ice, which makes finding food difficult, and are sometimes forced to decide between abandoning their chicks or starving themselves. In other years, when the ice breaks up closer to the coast, the colony thrives because there are so few birds that all benefit from the temporary surplus of food.

With around a quarter of a million birds, the Cape Crozier colony is bigger than many capital cities. The reason for this sprawl is a patch of open ocean in the sea

ice – a polynya – close to the colony. Ferocious katabatic winds blow out the ice early in the season and keep the area open for most of the year. Algae flourish and set off a localised but very productive food chain. As top predators, Adélies make use of the abundance and start breeding early, but in most years their advantage levels out later in the season once all pairs have growing chicks to feed.

There is one thing birds from all colonies share. They are fiercely loyal to their mates and breeding ground. Adélies mate for life and an established pair will usually set up their nest within a pebble's throw of last season's spot. Once chicks have fledged they spend five years at sea. If they survive long enough to start breeding, they invariably return to their natal colony. This loyalty makes Adélie penguins perfect study objects for evolutionary biologists.

By banding Adélie penguins (above left), scientists have been able to confirm that, though most birds are loyal to their natal colony, some move on to another group. Behavioural observations complement physiological and genetic studies of Adélies.
ANTARCTICA NZ PICTORIAL COLLECTION: J WAAS/K027 04/0

Ancient genes

Year after year, millions of penguins make the long march across sea ice to return to the same spot to start raising their chicks. During the breeding season, some adults and many chicks die, and so, over the millennia, the colonies also become ancient cemeteries. Today's breeding pairs are literally building their nests on top of the bones of their ancestors. Given Antarctica's cold and dry climate, the bones are well preserved and many still contain intact DNA.

The bleached bones are a treasure trove for David Lambert, an evolutionary biologist at Griffith University in Queensland. The way he tells the story, Lambert quite literally fell into his research. During one of his earlier Antarctic field seasons in 1984, he stumbled and, with his face planted on the ground, noticed bone fragments matted into the earth around the penguins' nests. The study of ancient DNA was in its infancy then, with only a few reports of traces of DNA having been found in museum specimens. But Lambert and his team soon realised that they had found a scientific asset. Since then, in a quest to observe evolution in the making, his team has dug deep into the frozen soils of several Adélie colonies to extract ancient DNA preserved in subfossil penguin bones, and to compare the genetic code with that of living birds. Excavating old bones and taking blood from irritable penguins are messy jobs but the subsequent analysis of DNA is a painstaking and sterile process that can only take place in a specialised laboratory. When I first met Lambert, he was based at Massey University in Palmerston North, New Zealand, where he had set up facilities to work with ancient DNA in close collaboration with University of Auckland biologist Craig Millar. The most ancient

Adélie colonies are one of the best sources of ancient DNA yet discovered and David Lambert's team has established the penguins as a model organism for the study of evolution. Based on his team's findings of a faster rate of evolution, Lambert has suggested that the 8 per cent genetic variance between the Antarctic lineage group and the Ross Sea lineage group of Adélie penguins arose over only 60,000 years. Previous estimates suggested that this change would have taken at least 200,000 years.

Lambert's group has already sequenced the complete genome of modern Adélie penguins and is now working on completing the task for ancient DNA from bones preserved since the late Pleistocene, a geological period when temperatures were 10 degrees warmer than today, to see if they find any differences.

COURTESY OF DAVID LAMBERT

Scientists take blood samples from an Adélie penguin at Cape Bird to extract DNA and compare it with ancient DNA from bones and frozen guano.
ANTARCTICA NZ PICTORIAL COLLECTION: WEI-HANG CHUA/ K030 03/04

bones his team had retrieved were more than 6400 years old. Once they had been ground up in a sterilised coffee grinder and digested chemically to release the DNA, the short fragments of ancient genetic code had to be amplified, sequenced, and matched with DNA from living penguins. It took a week to complete the sophisticated mathematical algorithms that use the divergence between the old and modern DNA to estimate the rate at which these sequences have evolved. The calculations produced surprising results.

Millar and Lambert worked on the assumption that the rate of change of the particular region of penguin genome they were using was about 20 per cent per million years. This figure was based on calibrations from the fossil record, the method of choice in lieu of ancient DNA. Palaeontologists estimate evolutionary rates by comparing the oldest-known fossils of two living species to determine the point in time when they developed from a common ancestor. Until recently, it was assumed that evolutionary change happened at a steady pace, like a clock ticking at standard intervals. This concept of a molecular clock was first introduced by Emile Zuckerkandl and Linus Pauling in 1962 when they noticed that the number of amino acid differences in haemoglobin between different lineages changed at roughly equal periods of time. They generalised their findings and proposed that the rate of evolutionary change of any protein was constant over time and over different lineages – a view that holds true for many, but not all, regions of the genome.

Lambert's Adélie bones revealed that the penguins' true evolutionary rate was 96 per cent per million years, or about five times faster than previously thought. However, the research also created a new dilemma. If all breeding birds returned to their natal colonies, there should be little interbreeding between the groups and, over time, they should become genetically distinct. Yet Lambert's findings showed no significant genetic difference between Adélies in different colonies, which implied that something was causing some of the penguins to break from their normal behaviour and explore new territory.

In March 2000, a giant iceberg known as B15 delivered an explanation. B15 remains one of the world's largest recorded icebergs. When it calved from the Ross Ice Shelf, it was nearly 300 kilometres (190 miles) long and 40 kilometres (25 miles) wide, with the trademark flat top that distinguishes Antarctic from Arctic icebergs. It ran aground and broke into several pieces. The largest fragment, B15A, remained stuck near Ross Island for several years, preventing currents from breaking up the sea ice in McMurdo Sound during summer. It doubled the distance icebreakers supplying the US research station had to cut through and Adélies were forced to walk considerably further between their breeding and feeding grounds.

When the iceberg finally detached itself, it collided with the Drygalski Ice Tongue, breaking off its tip and forcing a redrawing of Antarctic maps. As it moved further north, it broke into increasingly smaller fragments, some of which were spotted off the coast of the South Island of New Zealand in November 2006. That southern summer, the ocean opened up again north of the Cape Royds colony, restoring normal conditions for the penguins. But the population had taken a hit, with several failed breeding seasons and extremely low recruitment of new breeders.

David Ainley's team, working with Lambert's group, tracked nearly 10,000 birds that had been banded as chicks at the three Ross Island colonies. They found that the birds at Cape Royds had struggled to make the arduous trip home and that some had managed to immigrate to other colonies and interbreed with the locals. Genetically speaking, they were facilitating gene flow between the separate populations.

Consequently, Lambert believes that large icebergs are significant drivers of gene exchanges between colonies, given that most Adélie nesting areas lie close to ice shelves and glacier tongues, both of which calve off mega-icebergs. An estimated 200 such giant ice cubes have slipped off during the past 10,000 years and, typically, they will have stayed close to shore for much of their existence, just as B15A did, and are likely to have caused similar disruptions to ice conditions and penguin migration routes. The result of such upheaval is what biologists call micro-evolution – the same kind of small-scale genetic change within a species that has allowed pathogenic bacteria to become resistant to antibiotics.

Living by ice

Even without the hazards of massive icebergs, winter sea-ice extent has emerged as a good predictor of fluctuations in Adélie populations. An earlier study showed that in years of extensive winter sea ice, their numbers declined five years later. This lag period matches the time it takes a newly fledged chick to mature to breeding age and biologists interpret the statistic as a clear indication that many more young birds die during harsh winters.

Phil Lyver, a seabird ecologist at Landcare Research (like NIWA, one of New Zealand's independent Crown Research Institutes) who inherited the database when Peter Wilson retired in 2005, has continued to plot the Ross Sea Adélies' fate, from their most northern breeding site on the Balleny Islands to their southern outpost at Cape Royds, across 1200 kilometres (750 miles) in latitude. He found that since the start of the monitoring back in the 1960s, the birds have gone through a major population expansion. Initially, Adélie colonies throughout the region remained small but relatively stable. They began to grow quickly during the late 1970s until bird numbers had more than tripled a decade later, but then came a steep decline which lasted through much of the 1990s. Since then, the trend has been upwards again, but fluctuating widely, particularly at Cape Royds.

Adélie penguins are visual hunters and they need at least some daylight to forage. During the winter months, birds from all colonies travel about 4000 kilometres (2500 miles) with the pack ice to feed in the northern Ross Sea. But unlike whales and seals which can swim long distances to this winter feeding ground, penguins need additional transport. They float on chunks of ice to congregate in an area south of the unstable ice edge but north of the consolidated pack, at the edge of darkness. In some years, this narrow band of existence overlaps with the nutrient-rich waters. During harsher winters with more extensive sea ice the birds are carried too far north, while warmer seasons mean they never reach their destination. In both cases, many birds starve. Lyver conjectures that winter survival drives the birds' ecology and population dynamics.

Ross Sea Adélies live in the most southern part of the species' range. Further north, penguins nesting in colonies along the coast of the Antarctic Peninsula usually find enough food close by during winter and don't need to migrate far, but the southern birds often travel extraordinary distances between their summer breeding and winter feeding grounds. Their habitat has been in constant flux throughout the last six millennia, since the time of the first Egyptian pharaohs, during which the West Antarctic Ice Sheet withdrew southwards across the Ross Sea. As it moved, new potential breeding habitat became available as far as almost 78 degrees South. But to take advantage of the new sites, Adélies had to migrate.

David Ainley, an ecologist at H. T. Harvey & Associates in San Jose, California, thought that the birds' migration strategy could shed light on how Adélies coped with changing ice sheets in the past and how they might deal with fluctuating conditions in the future. At the end of the breeding season in 2003, he and his team picked several healthy birds at Cape Crozier and Cape Royds and fitted them with a geolocation sensor, just as they were heading off to their winter feeding grounds. They repeated the process for three years, a period which included both extensive and reduced winter sea ice. He found that the birds travelled an average of 12,760 kilometres (7923 miles) per year, partially aided by pack-ice movement, with a record round trip of more than 17,000 kilometres (10,600 miles) between autumn and spring.

The genome of Ross Sea Adélies differs from that of birds in more northern Antarctic regions, and Ainley thinks that is because they have been separated for long enough to become genetically distinct. That would mean that the southern penguin colonies have existed during the Last Glacial Maximum around 19,000 years ago, and that the birds have managed to survive as ice ages have come and gone. Nevertheless, Ainley is worried about the future. The rate of habitat change triggered by large-scale decreases of sea ice projected by future climate models may be unprecedented, and even if the penguins reach food-rich areas, they may not make it far enough north to enter the zone of twilight.

In 2010, the New Zealand and American penguin teams jointly published a paper predicting the penguins' future if Earth's average temperature were to rise by 2 degrees Celsius (3.6 degrees in Fahrenheit) above pre-industrial levels. Adélie colonies have vanished from the Antarctic Peninsula where this level of warming has occurred already. Since 1979, the western side of the peninsula has seen a 40 per cent decrease in the mean annual extent of sea ice, and biologists working at the smallest US Antarctic base, Palmer Station, now see the plight of the Adélie penguins nesting there as clear evidence for the impact of climate change. The teams studying penguins in the Ross Sea region expect a similar outcome, with three in four colonies decreasing or disappearing altogether.

One of the paper's co-authors is Gerald Kooyman, a physiologist at the Scripps Institution of Oceanography, who has dedicated his career to studying Antarctica's marine animals, including emperor penguins. He was the first to attach depth recorders to the birds and to track their foraging trips and 500-metre (1600-foot) dives. He expects that emperors will be doubly affected by future climate change because they need ice both to breed and to feed, and may lose half of their breeding colonies. While Adélies may be able to colonise new ice-free patches, emperors would struggle to find ice thick enough to support them through their winter breeding effort.

3
True Antarcticans

Leopard seals often hunt close to the beach at Adélie penguin colonies; the birds will hover by the edge of the water in groups, waiting for the first with enough pluck to jump in. ANTARCTICA NZ PICTORIAL COLLECTION: WEI-HANG CHUA/K030 03/04

The Antarctic seas are among the most hospitable in the
world . . . The waters are extremely rich in oxygen and swarm
with food, both vegetable and animal. Above, the turbulent,
stormy skies provide highways along which travel can be
astonishingly fast and almost effortless – for those creatures
that are sufficiently expert aeronauts. But if you are a land
animal, condemned to travel on foot, Antarctica is the most
hostile and forbidding place on earth.
– Sir David Attenborough, preface to *Life in the Freezer*

Penguins and seals are inextricably linked with the frozen landscapes of Antarctica, but their real home is the ocean. Although both come to breed on land or ice, they feed in the rich waters around Antarctica's coast and, with the exception of emperor penguins, escape north as soon as daylight begins to fade at the end of summer.

In many ways, the ocean provides a more pleasant environment. The water is frigid and always below zero (32 °F), but at least the cold temperature remains steady throughout the year. In contrast, temperatures on land can fluctuate dramatically between the seasons and even day by day, and so the true Antarctic survivors are a handful of species that live on land, stay all year round, and face the rigours of the long polar night. There are no land birds or mammals or reptiles, or any of the invertebrates familiar from other parts of the globe. Here, the largest animals are the size of a pinhead, sharing their habitat with a small number of creatures that are invisible to the naked eye. The highest form of terrestrial animal life in Antarctica is a wingless midge that grows to 12 millimetres (0.5 inches), and even that prefers to live in the more agreeable climate of the Antarctic Peninsula. Further inland, in what biologists describe as continental Antarctica, terrestrial life is mostly microscopic.

Over the course of evolution, animals and plants have developed two main survival strategies to cope with extreme conditions, including the freezing temperatures, lack of liquid water, dry air and high salinity that define much of the Antarctic environment. Some, such as emperor penguins, for example, have found a way around the problems and manage to survive and produce a new generation each year regardless. Others shut down and wait for better times, even if that means being active only for a few days each year and perhaps not reproducing for several years. The latter strategy, known as resistance adaptation, dominates among terrestrial life forms in Antarctica.

Continental Antarctica's land-based fauna is small in both size and numbers. It features worm-like nematodes that grow to a length of 1 or 2 millimetres (a little over half an inch), only slightly larger mites and springtails, and microscopic rotifers and tardigrades (aquatic invertebrates more colourfully known as wheel animals and water bears, respectively). Most of these creatures have evolved tricks to survive the harshest times in a state of extreme torpor. They save all their living for the few days when conditions are pleasant enough. Their optimal growth conditions often overlap with those of related species in more temperate regions, but when they encounter conditions that would kill others they simply shut down their metabolism. Usually, any creature that ceases all metabolic function is considered dead, but Antarctica's minute animals enter a dormant state on death's doorstep – a latent life.

Biologists term the process cryptobiosis, or hidden life. It differs from hibernation. While a hibernating animal can drop its metabolic rate by 80 per cent of resting levels, cryptobiosis means complete shutdown, with no measurable activity. Some animals can do it at any time. Others have survival or dispersal stages – cysts or eggs, for example – which act as lifeboats. The nematode record for latent life is 39 years, while rotifers have been revived from dried plant tissue after 120 years.

Herding worms

When Diana Wall arrived in Antarctica for her first field season in 1989, she thought she had landed on Mars. She was familiar with deserts, but the barren landscape of the McMurdo Dry Valleys seemed out of this world. Up to this point, the Colorado State University soil biologist had focused her research on the interactions between nematodes and plants in the arid regions of the American Southwest, but the complexity of these soil ecosystems prompted her to look for a pared-down version of a desert habitat. She found it in the sparse patches of exposed soil in continental Antarctica.

While a handful of soil collected elsewhere harbours hundreds of invertebrate species, those present in Antarctic soil can usually be counted on the fingers of one hand. Often, a few nematode species represent the top of a very simple food web, dining on bacteria, yeasts and algae.

Nematodes are ubiquitous, and more than 28,000 species have been described so far. Free-living nematodes have colonised almost every type of soil as well as freshwater and marine habitats. Parasitic species can affect most plants and animals, including humans. The best-studied nematode is *Caenorhabditis elegans*, a tiny, translucent roundworm found in many soils. It is one of the simplest animals with a nervous system and, in 1998, became the first multi-cellular organism to have its complete genome sequenced. Its entire body is made up of not much more than a thousand cells, each of which has been studied in detail. Thousands of papers have been published on this and other nematodes, all compiled in a set of data known as the WormBase.

Antarctic nematodes are equally simple in their body plan. They have an oesophagus, intestine, some muscles, and a minimal nervous system with no sensory organs. But their primitive makeup belies their ecological importance, the complexity of their lifestyle and the range of physiological adaptations they have evolved to cope with their environment. As Wall and her team began scouring the parched soils for microscopic life, they discovered that nematodes were managing to survive throughout the Dry Valleys, even in the most arid patches, despite the fact that they are essentially aquatic animals. The 'worm herders' also found that the nematodes' contribution to the turnover of organic matter in the soil was significant. In the short-grass steppe Wall studies at home, about 200 different soil invertebrates share the job of transforming decaying matter back into nutrients. In the Antarctic desert, two or three nematode species do the same work.

More than two decades after her first season, Wall's enthusiasm for her hardy research subjects and their icy home remains undiminished. Her focus now is on investigating the nematodes' capacity to cope with a changing environment, and to that end she set up a series of small experimental chambers within which she can manipulate temperature and moisture and monitor what happens underground.

In their natural Antarctic habitat, nematodes spend the biggest part of the year frozen solid and often almost completely dehydrated. They revive only during summer, when temperatures climb high enough to melt some snow or ice into a trickle of water, even if it is just for a few hours at a time. Like all terrestrial animals and plants in that continent, nematodes prefer ice-free specks of land, mostly dark volcanic soils, where summer temperatures can fluctuate between below zero and plus 25 degrees Celsius (32–77 °F). This means that they go through several freeze–thaw cycles each day, and the short bursts of activity between cold spells are their best chance at reproduction. To survive the long periods of winter dormancy and the shorter episodes of inactivity in summer, nematodes have evolved several tactics. Some dry out to avoid freezing, some tolerate freezing, some supercool, and some use a combination of all these methods.

Tough survivors

David Wharton's fascination with nematodes began when he was studying biology at the University of Bristol, taking a closer look at the structure of egg shells of a parasitic species. It turned out that nematode eggs are among the most resistant of any living structures. They survived everything Wharton tried out, including full immersion in sulphuric acid. A nematode egg is protected by a layer of lipids which restricts any exchanges with the outside world. Shielded within, the larva can survive total desiccation, freezing, and exposure to chemicals that would otherwise be fatal – including the fixatives that are used to kill and stabilise specimens for microscopy.

Even as a schoolboy, Wharton knew that he wanted to be a scientist. Inspired by classroom science experiments and natural history programmes, he developed an early fascination for creatures that remain invisible without a microscope and make their living as parasites in larger animals and plants. Initially, he wanted to study parasitic protozoa, a group of single-celled animals that includes amoebae, but a chance meeting with Bill Block, a pioneering scientist investigating cold tolerance in invertebrates at the British Antarctic Survey, got him wondering how nematodes survive extreme cold. In 1985, as a young and newlywed postdoctoral scientist at the University of Wales, he spotted an advertisement for a permanent research position in New Zealand. He applied, but was so sure he would never get the job, he didn't even mention it to his wife of six weeks.

Not much later, the Whartons were on their way to Dunedin, and a post at the University of Otago. The move heralded a shift in attention, away from parasitic nematodes to those that survive the hostile conditions of Antarctica. During his first field trip to the ice in 1988, Wharton collected several species and, eventually, managed to keep them alive in his laboratory, where he began testing their capacity to handle a range of environmental stresses.

When I visited Wharton's quarantine laboratory to have a look at his miniature menagerie through a microscope, I could see no discernible organs, and it was not even always clear which end was which. The most obvious anatomical feature was that some nematodes were packed with eggs. They were all females, and the eggs would eventually hatch into more females because this particular species reproduces without the need for fertilisation, in a process called parthenogenesis.

For Wharton, the nematodes' sex-free reproduction strategy provides a great advantage. He can breed an entire population from a single specimen of interest. During experiments, the worms live in Petri dishes filled with nutrient agar and a

good supply of soil bacteria to graze on, and are kept in an incubator at 20 degrees Celsius (68 °F). When not needed for research, they are stored in the freezer.

Perhaps one of the most remarkable nematodes is a species Wharton brought back from his first Antarctic visit. He isolated it from Ross Island, from an ice-free patch where loose soil and shingle are occasionally flushed in melt water from a snow bank – and it has survived in culture ever since. Surprisingly, this nematode grows best at 25 degrees Celsius (77 °F) and not at all below 6.8 degrees (44.2 °F). When it does get a chance, it can complete a life cycle from egg to adult and the next generation of eggs within seven days, but in its Antarctic habitat this may not happen for some years.

This nematode is a champion in a number of survival contests. It can avoid freezing by drying up almost completely, but its most outstanding ability is to tolerate total freezing, including that of the cytoplasm within each of its cells. Other freezing-tolerant invertebrates and insects stop just short of that. They can survive if the fluid that surrounds their cells freezes, but any intrusion of ice crystals into a cell is usually fatal. This particular nematode species was the first and so far only example of an animal that can bounce back from extensive intra-cellular freezing, seemingly unharmed.

It is such a fantastic feat that Wharton couldn't quite believe it himself at first, until he watched the process under the microscope. Nematodes are transparent and the progression of freezing is obvious because ice scatters light and the worm darkens as its body gradually turns solid. After analysing a video of a freezing

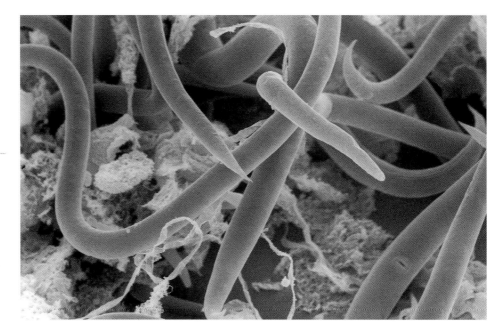

A scanning electron micrograph shows the Antarctic nematode *Panagrolaimus davidi*, which David Wharton has managed to keep alive in culture since his first Antarctic field trip in 1988. The nematodes are about 1 millimetre (0.04 inches) long. COURTESY OF DAVID WHARTON

nematode frame by frame, Wharton found that the first ice crystals enter through natural openings but then spread throughout the worm's body and into each of its cells very quickly, within seconds. More recently, he provided further proof that this nematode survives rapid intra-cellular freezing when he developed a technique to visualise the ice crystals within each cell. He froze nematodes, cooled them down further to minus 90 degrees Celsius (-130 °F) and then flushed them with methanol, which remains liquid at this temperature. The methanol replaces the ice crystals and, when the frozen specimen is inspected in an electron microscope, leaves behind large holes where ice had formed.

Wharton reckons that the rapid speed of ice formation within cells is one important factor in survival because the crystals have no time to become too big. But to emerge unharmed, the nematode has to keep the ice crystals in check even once it is completely frozen. It does so with the help of proteins that stop recrystallisation – which is what happens with anything that has remained frozen for some time, including ice cream that tends to turn slightly crunchy after a few weeks in the freezer. Even though ice formation starts with minuscule crystals, they continue to grow and then combine into larger chunks, which can cause considerable tissue damage. During summer, nematodes can freeze and thaw several times in a day. They don't have the time for repair pit stops between the cycles and instead rely on prophylactic damage control by keeping ice crystals small.

Worms in space

While he was still working on parasitic nematodes, David Wharton explored their capacity to live without water. Water is more important than food and most organisms don't last long without it. While some frogs can lose half of their body's water content, humans don't survive much more than a 10 per cent reduction. Invertebrates fare much better. Earthworms survive desiccation down to a fifth of their normal water content, but that is still nothing compared to the complete water loss found in some small spineless creatures, including nematodes.

Anhydrobiosis, or life without water, is another form of latent life, in this case triggered by extremely dry conditions rather than cold temperatures. The larvae of a certain midge species are the biggest organisms known to survive near-complete desiccation, but some nematodes are so good at it that Wharton and a colleague put them forward as promising candidates for space travel. The worms were booked on a Space Shuttle launch as part of a European Space Agency experiment in the mid-1980s, but in the end they didn't survive the preparatory training when they were put into an extreme vacuum.

They may have lost their seat on a space mission, but nematodes are nevertheless among the hardiest creatures when it comes to surviving water loss. They can last for several years in a state of reversible shutdown, reduced to tiny flakes of shrivelled-up tissue dust after losing more than 99 per cent of their body water. For Antarctic nematodes, desiccation tolerance adds another tool for dealing with the cold and avoiding the risk of tissue damage from ice crystals. A dried-up specimen has no water left to turn into ice, even if everything else around it is frozen.

Surviving the drying is only half the challenge, though. Water plays a crucial role in keeping cell membranes intact and supple, and without a protective mechanism, rehydration could cause as much tissue damage as thawing. As water molecules are sucked out during desiccation, membranes thicken to a gel. When water becomes available again, the gel rehydrates once more but leaks as it swells up – causing potentially fatal damage to cells.

The main protective agent is thought to be a simple sugar called trehalose, which replaces the water molecules to keep membranes more fluid and stop them from breaking. Once dry, trehalose turns into a glassy film, in a process similar to the melting of household sugar into caramelised sweets. The sticky sugar glass acts like a firm mould to hold tissue in place.

Most nematodes can produce trehalose in large quantities, which could explain why anhydrobiosis is widespread in this group. But some don't survive water loss despite the fact that they produce the sugar in response to desiccation stress, and Wharton expects that trehalose is part of a more general response to stress and only one of several, as yet unknown, adaptation strategies to life in extreme environments.

Compared with cold-tolerant insects, which have usually evolved one dominant strategy to cope with freezing temperatures, nematodes are more flexible and rely on several approaches, each depending on the specific conditions of their environment. Wharton's model nematode has at least four survival options. In a completely desiccated state, it avoids freezing and survives at minus 80 degrees Celsius (-112 °F) for a month or more. If it cannot avoid contact with ice and the crystals form rapidly, it survives complete freezing, including the interior of its cells, as described above. If it is exposed to temperatures just below zero (32 °F) and ice formation is slow, it dehydrates slowly as it cools and avoids freezing. And finally, as an embryo or larva, it is protected by an egg shell that allows it to cool down without freezing while encased in ice.

To freeze or not to freeze

The first descriptions of a complete organism that survived freezing go back to the mid-seventeenth century, when the British physician and experimenter Henry Power froze a jar of vinegar that had been infested with tiny worms, most likely nematodes. When unfrozen a few hours later, the worms revived as if nothing had happened. By 1683, Robert Boyle, who is often referred to as the founder of modern chemistry, also explored the effects of cold, inspired by the discovery that bodies that had been buried in the frozen soils of Greenland showed no signs of decomposition 30 years later. He tried freezing frogs and small fish, some of which survived short periods in ice, and was the first to show that meat and vegetables could be preserved by freezing, albeit at significant loss to their texture after thawing.

Before James Clark Ross ventured south to explore Antarctica, he joined his uncle John Ross on an expedition that located the magnetic North Pole – and found caterpillars that survived temperatures of more than minus 40 degrees Celsius (-40 °F). However, it wasn't until 1982 that scientists gathered conclusive evidence showing that some reptiles and amphibians and many insects indeed survive being partially frozen.

None of these animals can control their body temperature. They cool down to the temperature of their environment and have to find ways of either avoiding or coping with periods when it drops below freezing. Some insects avoid the cold by burrowing into the soil. Some, such as monarch butterflies, fly away to somewhere warm. Some simply freeze solid. New Zealand's alpine weta is a famous example of the latter approach. Its extra-cellular fluids can freeze completely, with about 80 per cent of its body water turning to ice. If a frozen weta is dropped, it will break. If it is allowed to thaw, it will walk off, unperturbed.

For larger species that either live in water or require moisture, and therefore cannot avoid contact with ice crystals, freezing tolerance may be the only strategy to cope with sub-zero temperatures (sub-32 °F). However, there is another option available to some of the tiniest creatures, as long as they can prevent ice from entering their body: supercooling.

A tiny droplet of water may not freeze until the temperature falls much lower than zero degrees Celsius (32 °F), the normal freezing point of fresh water. Freezing requires a trigger – a nucleator, such as a speck of dust. Water molecules themselves can become a nucleator if they group together and form the first tiny crystal, but such spontaneous aggregations are increasingly rare the smaller the volume and the warmer the temperature. Even the vibrations of clapping can snap freeze a droplet, but in the absence of anything that sets off the formation of ice, water will continue to cool down as a liquid. It will supercool – and this is how springtails survive in Antarctica.

When size matters

Few people can pinpoint the beginning of their career, but Brent Sinclair still remembers the moment that prompted him to embark on the study of Antarctic springtails. He was a third-year student at the University of Otago, working on freeze-tolerant alpine insects for his honours degree, when he heard a guest

This electron microscope image shows one of the springtails Brent Sinclair has studied in detail, a species called *Gomphiocephalus hodgsoni*. BRENT SINCLAIR AND MATTHEW DOWNES

Brent Sinclair analysing results in a tent laboratory, at Cape Hallett, in 2002. One of the experiments he conducted was to determine the springtails' supercooling limit by pushing them just beyond it and freezing them – which also kills them. He collected springtails at different times of the day, monitored the temperatures at which they froze and found that their capacity to survive the cold changed significantly, depending on whether they had just been feeding, as well as other factors that are not yet fully understood. COURTESY OF BRENT SINCLAIR

lecture by Bill Block. Block had spent his field seasons mostly in alpine or polar regions, much of it in Antarctica. Inspired, Sinclair abandoned his plans to look for postgraduate projects overseas, knocked on David Wharton's door and asked for a PhD topic that would get him to the frozen continent.

He chose springtails because so little was known about these tiny arthropods that even the question whether they were insects or not was an open debate. Whatever their precise taxonomic description, springtails are tiny wingless creatures, smaller than a grain of rice and just visible to the naked eye. They have a typical arthropod body plan, with a segmented body, jointed appendages and little antennae, and a velvety, wrinkled look. They get their name from a spring-like organ on the underside that allows them to jump, but in Antarctic species this is so reduced that their movement is more like a slow amble. Given that midges are only found at the most northern tip of the Antarctic Peninsula, Sinclair doesn't hesitate to describe them as the undisputed giants in the land fauna of continental Antarctica.

One of Sinclair's field seasons took him to Cape Bird for four months, during which he returned to Scott Base only once to have a shower and spent the rest of

A mass of springtails, or collembola, on a moss. They don't eat the plant itself but are thought to graze on microbes that thrive in the micro-habitat. Springtails have also evolved an unusual dispersal strategy – skin rafting. They have to shed their skin to grow and have been observed floating on the water-repelling tissue for up to six weeks. ANTARCTICA NZ PICTORIAL COLLECTION: CATHERINE BEARD/K024 04/05

his time turning over rocks in pursuit of his study objects. He found springtails everywhere, wriggly black spots among the pale grey pebbles.

Like most arthropods, springtails moult until they reach a final stage, usually their reproductive form, and the largest are generally the oldest. Their current cold survival record is minus 38 degrees Celsius (-36 °F), but that is just the lowest observed temperature. Clearly, they can survive an Antarctic winter. So far they haven't been found inland on the Polar Plateau, but even in their preferred habitat in ice-free coastal sites, winter temperatures can drop below minus 40 degrees. Springtails are often found in mosses, grazing on fungi, algae and bacterial mats that grow on the plant. One species Sinclair came across at Cape Hallett has piercing mouth parts and is thought to eat the eggs of other springtails and mites.

Without wings to fly away, no soil to dig into and nowhere else to go, Antarctic springtails have to put up with life in the freezer. Their Arctic relatives use cryoprotective dehydration – the same process some nematodes have evolved – and dry out as they cool down to lose any water that could freeze. But in Antarctica, springtails supercool.

Size matters in supercooling. Springtails are so tiny that the probability of spontaneous ice formation remains low. If they can avoid direct contact with ice, they can cool down to minus 20 degrees (-4 °F) without freezing. To make sure they last longer and can drop further down the temperature scale, springtails purge their gut to get rid of any potential nucleators and produce a cocktail of chemicals that prevent freezing. Glycerol, the same substance that has been used historically as an antifreeze in car engines, is one of the most common compounds found in springtails. Some produce trehalose and other carbohydrates, while others can make antifreeze proteins. So far only one such protein has been described chemically and it looks like nothing else found in other Antarctic species.

Yet, even though they inhabit one of the most forbidding natural environments, springtails don't hold the world record in supercooling. Sinclair, who now continues his work on cold tolerance in insects at the University of Western Ontario, says that particular credit goes to a small beetle that lives in the interior of Alaska and can remain unfrozen down to minus 100 degrees Celsius (-148 °F), thanks to glycerol.

Antarctica's forests

Plant life in Antarctica is mostly two-dimensional. With the exception of two flowering plants that grow only along the western side of the Antarctic Peninsula and some of its offshore islands, all other species prefer to hug the ground closely. Even in maritime Antarctica with its milder winters, summer rains and several weeks of above-zero temperatures, the most common plant communities are undulating carpets of mosses and lichens, only occasionally interspersed with swards of the taller Antarctic hairgrass and the flowering stalks of the Antarctic pearlwort. It is a luxuriant display compared with anything that survives in continental Antarctica. The Antarctic Peninsula accommodates about 350 different lichens, more than 100 mosses and 27 liverworts, but the numbers drop quickly along the coast of the continent and further inland, where only a third of the lichen species, 20 mosses and a single liverwort manage to eke out an existence.

University of Waikato plant physiologist Allan Green still finds plenty of material to study. Some of his inspiration about where to look for patches of interesting botany has come from the early explorers. Waiting for good weather is part and parcel of Antarctic science, and whenever Green was stuck at Scott Base waiting for the next helicopter flight out into the field, he would bide his time poring over old books in the library. The earliest descriptions of Antarctic plants go back

to Joseph Dalton Hooker, surgeon-cum-naturalist on Captain Ross's nineteenth-century voyage of the Southern Ocean. Although earlier expeditions had described and collected some plant specimens, Hooker was the first professional botanist to make a comprehensive collection of plants throughout the expedition. On his return to Britain in 1844, he promptly began publishing several volumes on the plants discovered en route. His list of Antarctic flora was short: nine lichens, five mosses and four algae.

Green was more encouraged by the diaries of Griffith Taylor, one of Scott's men who led a geological expedition to Granite Harbour in November 1911, not long after Scott himself had departed on his fatal journey to the South Pole. Taylor's party set up camp on a beach of granite boulders in a bay they named Botany Bay for its proliferation of mosses and lichens. They found such rich growths that they even used dried moss to fill in cracks in the walls of their rock shelter.

Botany Bay became Green's favourite natural laboratory for many years. Even after extensive exploration of other areas, this verdant patch at 77 degrees South remains the richest site in the southern Ross Sea region. It represents the

TRUE ANTARCTICANS   »   135

southern limit for liverworts and harbours the only species to survive in continental Antarctica as well as several mosses and more than 30 species of lichens, some of which grow to more than 20 centimetres (8 inches) in size.

Strictly speaking, lichens are not plants, but a composite between a fungus and an alga living in a symbiotic relationship. The alga is the true plant component. Like all plants, it can convert carbon from the air into larger sugar molecules through the process of photosynthesis. It feeds the fungus in exchange for shelter and transport – and as a result of this cohabitation, lichens have been able to spread across the globe to some of the most inhospitable places. The symbiosis has proved so successful as a guaranteed food supply for fungi that one in five species have adopted it – and taxonomists prefer to group lichens within the kingdom of fungi. They also name lichens after the host fungus, which largely determines its growth form and size. However, in Antarctica lichens are considered part of the terrestrial vegetation because of their photosynthesising lifestyle and prominent role in the sparse botanical niches.

Whatever their taxonomic pedigree, lichens are ecological endurance athletes. During his first summers on ice back in the early 1980s, Green remembers thinking that he would eventually discover that lichens had developed some extraordinary adaptation to the extremes of the Antarctic environment. He still hasn't found it.

Instead, he gradually learned that lichens in Antarctica respond to temperature, light and water stress in ways that are similar to those of their widely spread relatives elsewhere. The main difference is that Antarctic lichens cope better with waiting. Green thinks the one reason for the almost tenfold difference in the number of lichen species living on the Antarctic Peninsula as opposed to the icy interior of the continent is that a select group can handle very low activity and long periods of dormancy. Although the basic underlying physiology is similar for all Antarctic lichens, some focus their effort on survival rather than growth, and spend most of their time in a state of latent freeze-dried potential, waiting for conditions to improve. They grow so slowly that it took more than two decades for Green to detect any significant gains in the lichens he first sampled and measured during his early fieldwork. With an average growth rate of 1 millimetre (0.04 inches) per century, any lichen patch that is bigger than a quarter could well be a few thousand years old.

Like the continent's terrestrial animals, all Antarctic plants spend most of their time in shutdown mode. During the perpetual darkness of winter, without light or liquid water, all metabolism ceases. Lichens and mosses have no roots and no capacity to store water. They rely directly on drawing moisture from their environment, and as a result they dry out, and remain dormant, during periods when all water around them has turned to ice. As soon as some of it melts, the plants sponge up the liquid and start up again. In contrast to desiccation-tolerant animals which produce sugars to prevent tissue damage during rehydration, lichens and mosses depend on a suite of proteins. This strategy is typical for these more primitive plant groups throughout the world, but Green believes that Antarctic species are particularly robust and able to recover more quickly – with the added advantage that desiccation also protects them from the extreme cold by withdrawing water before it can freeze. Lichens can survive several years in a dried state. Mosses manage several decades.

Perhaps the one ability that makes lichens good candidates for Antarctic survival, and distinguishes them from mosses and liverworts, is their capacity to photosynthesise at temperatures well below zero (32 °F). It is not a skill exclusive to Antarctic species, though. The record goes to a temperate species which was still operating, albeit at a low level, at minus 24 degrees (-11.2 °F) in the laboratory. In the field, Green found that lichens absorb the sun's energy so well that their shoots can be several degrees warmer than the ambient air, allowing them to go about their business while everything around them is still frozen. With this jump-start, lichens get a longer growing season than all other plants – and animals – on the continent.

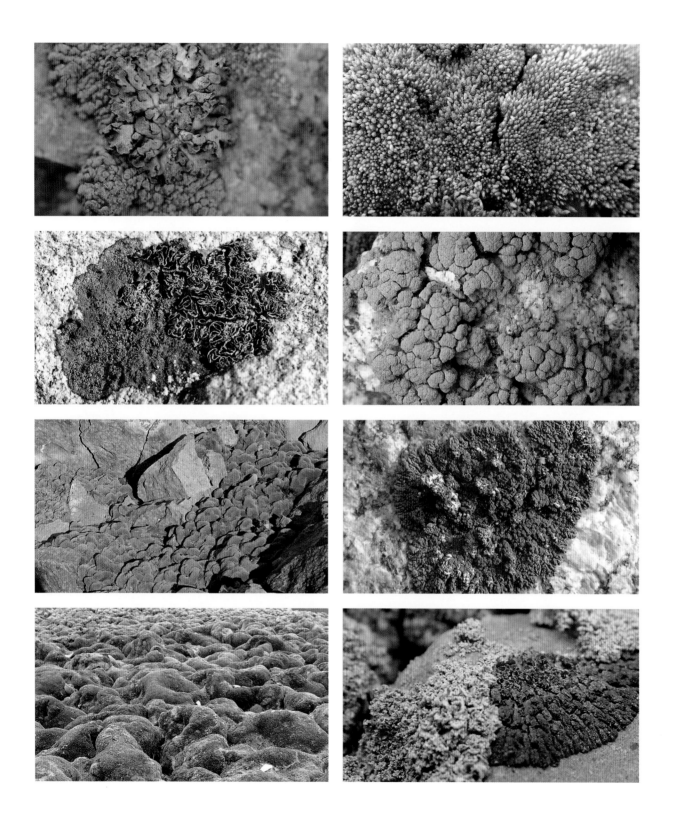

In many ways, extreme cold and lack of water are minor problems for lichens. In a desiccated state, all Antarctic lichens survive being dropped into liquid nitrogen at minus 196 degrees Celsius (-320.8 °F), and one species recovered after almost four years at minus 60 degrees (-76 °F). Their biggest stress comes from the intensity of light during the short summer. In more temperate regions, lichens are most active early in the morning when the light is still dull, or on rainy and overcast days. In Antarctica they don't have such options. They need light and liquid water to operate, so their only chance is to make use of the periods of full sunshine. That is why Antarctic lichens are things of beauty, with some of the strongest displays of colour seen on land. They range from the more conventional greens and browns to bright red and jet black, all in an effort to filter out most of the light that burns down on them.

The next biggest stress factor is the ultra-short period of activity. Despite their capacity to get started at freezing temperatures, some lichens are only active three or four times a year, for a few days or even hours each time. They operate in a mode of minimal expenditure that makes their reserves last longer but allows for not much else. Nevertheless, lichens have been found as far as 86 degrees South – only 400 kilometres (250 miles) from the pole – and Green has explored an area at the Beardmore Glacier, at 84 degrees South, which almost rivals Botany Bay in terms of its botanical diversity. He discovered a description of this spectacular site in a 1960 issue of the *New Zealand Alpine Journal*, which featured an account of a geological survey by one of the participating students. More than four decades later, Green sent a field party to explore this improbable icy garden. It found 26 lichen species, almost as many as in Botany Bay, but the most intriguing discovery was that a quarter of the species had never been seen anywhere else in continental Antarctica. The nearest other site to accommodate such an array was on the Antarctic Peninsula and then again on King George Island in the South Shetlands. Green's team had stumbled upon a tiny relic population, left over from a time before the Ross Ice Shelf spread out across the vast area it covers today.

Lichens only have to be near water but mosses grow in it. They snuggle into any cracks and crevices where melt water pools, or grow where water runs across the ground, however rarely. Survival can be a trade-off between closeness to water and shelter from the wind. Wind is both a friend and foe. It distributes freeze-dried patches of moss to potentially new habitats but it also sandblasts sensitive new shoots, often eradicating the entire plant. In some places, mosses survive only by growing beneath the surface to escape the abrasive power of Antarctica's storms.

Like lichens, mosses find the warmer and wetter climate of the Antarctic Peninsula more liveable and diversity declines dramatically in continental

The forests of Antarctica: Antarctic lichens and mosses produce brightly coloured pigments to filter the intense sunlight that burns down on them during the short southern summer. Some lichens form small leaves (foliose lichens), while others cover the ground like a crust (crustose lichens). They all make the most of any protection from the elements they can get. OPPOSITE, CLOCKWISE FROM TOP LEFT *Physcia caesia* and *Caloplaca citrina*, two species of Antarctic lichen that are also found in temperate regions; a bed of *Bryum argenteum*, one of the most comprehensively studied of Antarctic mosses; *Candelariella flava* is a lichen that likes nitrogen and usually grows near penguin colonies; *Xanthoria elegans* is more commonly known as the elegant sunburst lichen; *Buellia frigida* with a species from the orange-hued genus *Caloplaca*, found throughout the world; a patch of *Bryum pseudotrichetrum* at Cape Hallett; a bed of *Bryum argenteum*; and *Umbilicaria decussata* and *Buellia frigida* lichens, freeze dried. ROD SEPPELT/ AUSTRALIAN ANTARCTIC DIVISION; ANTARCTICA NZ PICTORIAL COLLECTION: TRACEY JONES/K024C 06/07; CATHERINE BEARD/K024 04/05

Antarctica. Of 111 Antarctic mosses, fewer than ten species survive in the Ross Sea area and only three make it further south than Ross Island.

Mosses also share the lichens' dilemma of having to be active at the height of the southern summer, when the sun never sets, and have evolved an arsenal of pigments to shield themselves from the intensity of solar radiation, including some that block out ultraviolet radiation. They grow faster than lichens and form clearly visible clumps, but compared with plants from temperate regions, Antarctic mosses are nevertheless slow. When Green revisited moss patches he had studied two decades earlier, he could still see where he had taken samples for his measurements. But he also found that mosses can speed up considerably if they need to. His team shaded some exposed plants until they turned bright green. When they were stripped of their protective cover, the plants took less than a week to change colour and regain their full ultraviolet protection – not by producing and redistributing more protective pigment but by pushing out new yellow-green shoots.

For mosses in Antarctica, sex represents the biggest waste of energy. Elsewhere, these primitive plants produce fruiting capsules to spread their spores, and some try to do the same on ice, invariably without luck. Cape Chocolate, south of the Ferrar Glacier, is the southernmost place where botanists have found capsules on some plants, but none managed to mature enough to produce spores. The brief sprints of activity are simply not sufficient to complete a cycle of sexual reproduction, and instead Antarctic mosses rely purely on vegetative propagation – small, dry fragments being blown about by the wind.

One of the most comprehensively studied Antarctic mosses is a cosmopolitan species which grows on almost any footpath – *Bryum argenteum*. The genetics of its Antarctic form, which can respond quickly to changes in light and ultraviolet radiation, suggest that it arrived before the large ice sheets, and that the current patches are relics of a much larger and more widely spread ancient population. Green reckons the same might apply to all mosses that survive in Antarctica today, as their capacity to sit out hostile conditions makes it more likely that they have a long history on the continent.

Fungal decay in the freezer

Captain Lawrence 'Titus' Oates is best known for the nobility of his last words: 'I am just going outside and may be some time.' Weakened by frostbite and sheer exhaustion and aware that he was slowing down his companions, he walked out of the tent into a blizzard to give Scott, Bowers and Wilson a better chance as they struggled back from the South Pole.

Oates is also immortalised in an image of 'the tenements', the tightly fitted bunks in an alcove of the Terra Nova hut, taken by Herbert Ponting a few weeks before the start of the ill-fated march south. His bunk is easy to identify. Bridles and ropes hang from one corner, ready to be fitted on the ponies that had sailed south with the expedition. Oates had been selected for his horsemanship and his main task was to keep the ponies in good shape. Indeed, he was so concerned about them that he convinced Scott to load an extra 15 tonnes (16.5 tons) of hay as the *Terra Nova* lay in Lyttelton for final preparations, only to then complain that it was too mouldy and not good enough for his precious animals.

Almost a century later, Waikato University biochemist Roberta Farrell proved Oates right when her team found the hay bales that remain at Scott's hut at Cape Evans full of fungi, which were still alive and proliferating. Farrell had come to inspect the historic huts for signs of deterioration and to understand whether it was caused by the environment or microbes. She discovered thriving populations of several types of fungi, including the species that turns blue cheese into a delicacy. Some of the organisms she found seemed to have made their way to Antarctica as stowaways with the early explorers or modern tourists and then managed to survive. However, her most extraordinary find was a fungus that is native to Antarctica, has probably survived for millions of years, and is now slowly gnawing away at the huts' timber.

Fungi are so different from other organisms that they have been given their own taxonomic kingdom, equal in ranking to animals and plants. Most people only get to see them as mould on spoilt food or as mushrooms, the fungal fruiting bodies that emerge above ground to release millions of spores. But fungi are cosmopolitan, and their lineage also includes rusts, smuts, puffballs, truffles and yeasts, as well as many other less well-known groups.

The living portion of a fungus is both huge and microscopic at the same time. Each individual thread-like cell is invisible but they are interconnected to form a huge network, the mycelium, and this filamentous growth can penetrate vast patches of ground. Fungi cannot ingest their food like animals, and neither can they make their own like plants. Instead, they feed by absorption, and to do that they have to grow into their substrate – some form of dead or live organic material – and release digestive enzymes to break it down.

American-born Farrell is a biochemist with a passion for fungi. She spent the first part of her career working for an international pharmaceutical and chemical company, scouring fungi for potentially useful enzymes and applications. With a focus on the pulp and paper industry, she extracted a fungal enzyme to help decrease the amount of chlorine used in the paper-bleaching process and used a fungus to develop a biological solution to the long-standing problem of resin residues in paper. When she moved to New Zealand in 1995, she continued to work with fungi and on similar problems, but Antarctica beckoned with the promise of discovery.

During her first visit in 1996/97, Farrell's initial goal was to establish whether this continent that had once been cloaked in trees still harboured any wood-decaying fungi at all, or whether they had all disappeared with the forests. During a chance meeting with a team of conservators at the Scott Base canteen she learned about Antarctica's historic huts and the efforts to protect them from degradation.

Back then, the New Zealand Antarctic Heritage Trust was sending a small team to the ice each year to carry out basic maintenance work at four huts built by Scott, Shackleton and Borchgrevink for their respective expeditions. Each of these pre-fabricated huts is a time warp without parallel. Between them, they contain several thousands of artefacts, from boxes of 'Homelight' lamp oil to oozing layers of seal blubber and an emperor penguin carcass on Scott's desk, still awaiting close examination a century later.

In 2002, the trust launched a major restoration project to preserve the huts and to limit further degradation, but when Farrell first explored these shrines to the Heroic Era, there was little scientific understanding of how well they had stood up to the elements over the decades. She took some samples from Scott's Terra Nova hut and contacted former colleague and friend Bob Blanchette at the University of Minnesota, with whom she had already collaborated on fungal applications for the pulp and paper industry. Blanchette had also investigated fungal decay in totem poles and other historic structures. This was the start of an ongoing collaboration in Antarctica.

Initially, Farrell and her team found mostly non-biological damage caused by wind, ice and salt. Salt gets into the wood as it would in any house built on the coast, but in Antarctica, rain never washes it out again. As a consequence, it accumulates in the timber and soil to concentrations that can be several thousand times higher than elsewhere, corroding surfaces and eroding the interior structure of the wood.

Then, the team dug down into the soil to expose the buried end of a board and unearthed clear traces of fungal decay. It took them by surprise. Usually, wood decay happens only in moist and warm conditions. Although ice had crept under the floorboards and built up around the outside of the hut, increasing moisture levels inside, the average temperature hovered around minus 20 degrees Celsius (-4 °F) and should have been far too cold for any fungal rot. The extreme cold has

indeed protected the huts to a degree, but Farrell estimates that over the century since they have been erected, there was an aggregate period of somewhere between a few months and four years when conditions were good enough for fungi to grow and cause damage.

The fungus the team identified delivered another surprise. It belongs to a genus of soft rot that has also turned up in the Arctic and cold deserts in western China but is not usually a dominant part of the fungal flora. Several species of the genus, called *Cadophora*, appear to be unique to Antarctica and Farrell proposes they have survived as a relic from times when Antarctica was warm and green and wood decay fungi flourished in coastal swamps and peat bogs.

Elsewhere, soft rot fungi grow slowly and don't cause much damage, but inside the huts this fungus has found a special niche, probably thanks to the lack of competition from other types of decay fungi which didn't manage to survive. Farrell and her team found that in some places the fungus has caused considerable

Antarctic Heritage Trust conservators working in the physics area of Scott's Terra Nova hut, at Cape Evans. Scott used this hut during his final expedition, officially known as the British Antarctic Expedition 1910–13. The Antarctic Heritage Trust began work on this site in 2008 and conservators have since excavated the remains of Bowers' Annex from underneath snow and ice and installed five vortex generators to the south of the building to reduce the build-up of snow in the future. AHT/J. STEFAN

damage, eating the timber inside out. It grows its microscopic filaments right into the timber, boring tiny holes into each wood cell and pouring its digestive enzymes inside. Viewed through a microscope, the end result looks like miniature chicken wire – a mesh of empty, perforated cells that have lost any structural support.

When the Antarctic Heritage Trust began its restoration project in 2003, the scientists joined the conservators as they lifted floorboards and wall panels to remove ice that had built up underneath the historic huts. They took samples of fragmented wood, surfaces and artefacts that had not been exposed for a century. Up to that point they had expected to find a few thriving fungal specialists but a low diversity in species, confirming what most other people working on microscopic organisms in Antarctica thought at the time. Farrell threw that view out in the first season when the team discovered more than 120 new fungi.

One winter, Farrell travelled to Antarctica on Winfly (Operation Winter Fly-In), the first flight to land on the ice runway in August, just before the sun emerges

above the horizon again. The huts had been closed since February and she was curious to see how the fungi had made it through the winter. Wrapped in survival gear that covered everything except her eyes, sealed in a protective suit and with a small headlamp as the only source of light, she took samples in Ponting's studio at minus 40 degrees – and captured hundreds of viable fungal spores. Clearly, the fungi survived and even managed to reproduce during the depths of winter.

Intrigued, Farrell set out to explore fungal diversity in pristine sites. She wanted a baseline to assess whether the huts had been colonised by fungi that had mostly arrived with them or had been there all along, just waiting for better times. The team found that the same soft rot that was slowly eroding the timber in the huts also made up about a tenth of the fungal flora in sites far away from any external source of food.

When Farrell's team put it through its paces in the laboratory, it grew as happily at plus 20 degrees (68 °F) as it did at zero (32 °F), suggesting that it was indeed a

The interior of Scott's Terra Nova hut. As part of the hut's restoration project, conservation carpenters have reinstated the building's internal division known as the 'bulkhead', returning the space to a historically accurate layout which clearly delineates divisions between the officers, the scientists and the men. AHT/A. FASTIER

During the summer of 2010, Roberta Farrell's team joined forces with two companies, Geometria and Archifact, to image Scott's and Shackleton's huts to an unprecedented resolution, using 3D laser and light scanning technologies. This image shows Scott's Terra Nova hut, with the vortex generators installed by Antarctic Heritage Trust conservators in the background. These laser images are a vast improvement over traditional surveying methods, and the team hopes to collect further scans over successive years to monitor seasonal deformation by snow and ice.
GEOMETRIA

survivor from a time when Antarctica was warmer and forested. Her focus for the next research seasons will be to trace the emergence of the Antarctic *Cadophora* species on a geological timescale to understand the past and get a better idea of how the fungus might proliferate in the future.

Hot on ice

Late in 1980, a helicopter dropped off Hugh Morgan and Roy Daniel on a ridge that separates the old and new craters of Mount Erebus. They had come to begin their altitude acclimatisation for their later field trip to the summit of the world's southernmost active volcano where they planned to look for microbes that like it hot.

This was only a few years after the first such heat-loving, or thermophile, bacterium had been discovered, and their quest to find them in Antarctica had all the excitement of an expedition to uncharted territory. Morgan had already pioneered techniques to keep thermophile bacteria alive in laboratory conditions when he isolated the first such organism from a hot spring in New Zealand. Both had worked in geothermal areas in Iceland and the US and had helped to establish the thermophile research unit at Waikato University. They felt like they were part of a global race to find the microbe with the highest optimum growth temperature and to push out the boundaries of where life was thought to be possible.

The hot seeps on Mount Erebus are different from other geothermal areas. There are no hot pools or springs. Instead, jets of hot steam rise from the deep, condense on the surface, and freeze into tall, hollow chimneys of ice, still fuming. As small trickles of water seep back into the warm soil, algae and mosses grow on the moist surface, adding to the flow of nutrients that feed bacteria in the layers below.

As a microbiologist, Morgan was keen to see which types of microbes they would find, while Daniel, a biochemist, had his sights on the enzymes within these

On the slopes of Mount Erebus, jets of hot steam rise from fumaroles and condense to form cavernous grottos and tall chimneys of ice. ROB McPHAIL

Craig Cary taking soil samples at Tramway Ridge, near the summit of Mount Erebus. Cary initially focused his genetic studies on the microbial communities that survive in the arid soils of the McMurdo Dry Valleys (see pages 166–174), but recently he has revisited sites that were first explored by Roy Daniel and Hugh Morgan. By using DNA fingerprinting techniques, he can describe microbes that live happily in the warm volcanic soils but would not survive in a culture medium. The study of heat-loving, or thermophile, microbes continues to challenge our understanding of the upper limit of microbial life and how life may have originated on Earth. Tramway Ridge is the southernmost geothermal area on the planet, an ideal environment to study thermal adaptation in microbes. ANTARCTICA NZ PICTORIAL COLLECTION: CRAIG CARY/K023 05/06

A team of microbiologists measuring soil temperature and taking samples on the slopes of Mount Erebus.
ANTARCTICA NZ PICTORIAL COLLECTION: CRAIG CARY/ K023 05/06

organisms. Specifically, he was looking for new proteases – enzymes that break down proteins – to compare with their equivalent molecules from other thermophiles and more conventional microbes. Proteases were already reasonably well understood at the time. They are one of the first enzymes to have been studied in detail, with experimentation going back to the Victorian era when dog and pigeon faeces, full of proteases, were used to tan leather.

Proteases are ubiquitous. All living organisms have a suite of these enzymes to aid the digestion of protein-rich foods. Daniel could see the potential for new commercial applications, such as in cleaning agents and washing powders that work best at higher temperatures.

Once the team arrived at their summit camp, only a few yards below the crater rim, they began looking for the hottest spot by sticking thermometers into patches

Fumaroles and ice chimneys below the crater rim of Mount Erebus. ROB McPHAIL

of bare soil. Hundreds of soil samples were taken and tested for microbes that could be purified into culture. As it turned out, the most successful microbe was the first they managed to cultivate – a Bacillus species named simply Antarctica Erebus One, or AE1. Its protease showed two desirable characteristics that made it a promising candidate for commercial success: it remained stable at high temperatures and its activity could be turned on and off by changes in the concentration of calcium. AE1 and its enzyme soon became the foundation stone for a new company and the essential ingredient in a DNA analysis kit designed to speed up the processing of forensic samples.

As Morgan and Daniel were collecting their samples, they were aware that they were not seeing the full picture. This was long before DNA fingerprinting techniques were cheap and quick enough to make them useful tools in the field. The

pair had to rely on culturing techniques and microscopy, most of it carried out back in the laboratory in New Zealand, to extract and identify their organisms. They knew that for each one that could be taken into culture there were likely to be many more that couldn't.

Craig Cary and Rob Kirkwood, dressed in white overalls to avoid contaminating their soil samples, explore Tramway Ridge, near the summit of Mount Erebus. ROB McPHAIL

Bacteria are, and always have been, the dominant forms of life on Earth. The fossil record of life begins with bacteria some 3.5 billion years ago and they remained the planet's sole occupants for half of life's history until the first more complex cells began to appear about 1.8 billion years ago.

Before DNA sequencing provided a means for mapping evolutionary relationships, bacteria were usually all dumped into one big taxonomic group and distinguished mostly by their shape. Genetics opened a window on the complexity of this group and showed that, despite the lack of anatomical diversity, bacteria

Craig Cary and Rob
Kirkwood on Tramwa[y]
Ridge. ROB McPHAIL

are more diverse genetically than all three kingdoms of multi-cellular organisms – plants, animals, and fungi – combined.

Bacteria inhabit every place that can support life – and in astronomical numbers. One of the most widely known statistics is that the number of *E. coli* that live in the gut of one person during his or her lifetime far exceeds the total number of people who live now and have ever lived. There are millions of bacteria in a drop of saliva, and hundreds of thousands on each square centimetre of human skin.

The temperature tolerance of bacteria exceeds that of any other known organism and the fringes of life's extremes are usually exclusively bacterial. Together with fungi, bacteria are the main decomposers of dead organic material, recycling it back into a usable form of nutrients, but the range of metabolic tricks they have evolved to make use of almost any substrate leaves every other taxonomic group in their wake.

At the time Morgan and Daniel went looking for thermophiles in Antarctica, this group was the best-known example of extreme survival on land and in the ocean. Since then, bacteria have been discovered, and cultured, from oil drillings and other places deep beneath the ocean or land, drawing their energy not from the sun but from the Earth's interior. They have been found thriving in extremely saline, acidic or alkaline environments, under extreme pressure and extreme temperatures at either end of the scale.

Bacteria also hold the record in latent life. They have been revived from canned food that had remained sealed for more than a century and from a 166-year-old bottle of Porter ale that had sunk to the bottom of the ocean with a wrecked sailing barge. Bacillus spores sprouted new life after being preserved with a bee in amber for 20 million years, and the same type of bacteria is thought to have survived 250 million years in tiny bubbles of liquid in salt crystals.

Morgan and Daniel regard the Earth's interior as the latest, but probably not the last, frontier for bacterial discovery. This subterranean microflora, living anywhere between 2 and 5 kilometres (1.2 to 3 miles) below the Earth's surface, is invariably bacterial and thermophile and extends the belt of life to a considerable depth beyond what was once accepted as Earth's biosphere.

At the cold end of the temperature scale, microbiologists continue to make surprising discoveries in some of Antarctica's most extreme environments.

During the 2010/11 summer season, New Zealand scientists joined American colleagues on the Beardmore Glacier for one of the largest deep-field camps in Antarctic history. Discovered by Shackleton in 1908 during his Nimrod Antarctic expedition, the Beardmore, more than 700 kilometres (430 miles) south of McMurdo Station, was Scott's preferred route to the South Pole in 1911. The remote camp became a temporary home for 75 people, including a New Zealand contingent of geologists and biologists. Tim Naish led a geology group to explore an area known as the Cloudmaker, where glacial sediments have recorded oscillations in the margin of the East Antarctic Ice Sheet. Freshwater ecologist Ian Hogg braved icy winds to look for life on Mount Howe, the southernmost exposed rock on the planet, some 300 kilometres (190 miles) from the South Pole. While lichens have been found as far as 86 degrees South, Hogg found that Mount Howe, at 87.2 degrees South, may not support any life other than microbes. However, in other areas further north, he discovered several new populations of springtails and mites as well as lichens. Hogg's approach is to study the entire landscape to get a better understanding of the factors that control the distribution of invertebrate animals and plants in Antarctica – and eventually to predict where life may exist, even if only in small pockets.

This image shows Ian Hogg's team, from left to right: geomorphologist Bryan Storey (University of Canterbury), microbiologist Hugh Morgan (University of Waikato), Jason Watson (Antarctica New Zealand, field safety), lichen and moss specialist Leo Sancho (Universidad Complutense Madrid) and Ian Hogg (University of Waikato). CTAM stands for Central Transantarctic Mountains, some of which are visible in the background. COURTESY OF IAN HOGG

4
Oasis in a Frozen Desert

One of the glaciers of the McMurdo Dry Valleys. The Taylor Glacier flows for more than
50 kilometres (30 miles) from the Polar Plateau into the western end of the Taylor Valley.
ROB McPHAIL

I cannot but think that this valley is a wonderful place.
We have seen to-day all the indications of colossal ice action
and considerable water action, and yet neither of these agents
is now at work. It is worthy of record, too, that we have seen
no living thing, not even a moss or lichen; all that we did find,
far inland amongst the moraine heaps, was the skeleton of a
Weddell seal, and how that came there is beyond guessing.
It is certainly a valley of the dead.

– Robert Falcon Scott, *The Voyage of the 'Discovery'*

On October 12, 1903, Scott and eleven of his men set out to explore the interior of Victoria Land and the mountains further inland. Having pushed further south than anybody else during the previous summer, albeit without leaving the Ross Ice Shelf, then known as the Barrier, this foray to the western mountains was the Discovery expedition's second main journey. Scott's hope was to penetrate far into the interior of the continent.

One of his party's first challenges was to scale a massive glacier, named after the expedition's geologist Ferrar. They travelled fast, but only six days later Scott ordered the return to the ship when it became clear that the harsh ice was shredding the silver coating on the sledge runners. The only intact sledge, packed with everything except the provisions the party would need for the return trip, was left behind in the glacier basin. A fortnight later, a smaller party embarked on a second attempt, only to discover that a violent storm had ripped open an instrument box on the sledge and blown away a small volume called *Hints to Travellers*. It contained all the information and logarithmic tables needed to navigate by the sun and stars in the featureless landscape beyond the glacier.

Scott decided to push on regardless. On November 4, the next gale struck and the men spent a week sheltering in their canvas tents, spending 22 out of every 24 hours wrapped up in their sleeping bags, unable to sleep, enduring. Their perseverance paid off. Not long after the storm, Scott's men became the first to walk on the Polar Plateau, the vast expanse of flat ice that presses down on the continent so heavily that it dents the planet. They continued man-hauling their sledges westward across the plain, with no landmarks in sight, for another two weeks, but it was clear that some were struggling to keep up. Scott decided to send the weaker men

back and to continue west with only two of his companions, William Lashly and Edgar Evans. When their compass needle began pointing south, they knew that they were travelling south of the south magnetic pole, but without their navigational tables they could not be sure exactly how far they had gone.

The return journey was marked by unease. Scott's rule-of-thumb navigation provided no certainty that they would find the summit of the Ferrar Glacier again, and low-lying cloud did little to alleviate the tension. Hungry and desperate to get back to their ship, the three men continued to haul their sledge due east despite the snow drift. The weather was so bad that they only realised they had reached the top of the glacier when the sledge overtook them on the steepening slope. Then, Scott and Evans nearly fell to their deaths in a crevasse, but their narrow escape boosted their spirits and energy, and instead of continuing down the length of the ice tongue, they decided to explore the northern arm of the glacier. Suddenly, the ice receded and the view opened across a 'curious valley' of sandy stretches and jumbled boulders, completely free of snow and, according to Lashly, 'a splendid place for growing spuds'. They had discovered one of Antarctica's rarest phenomena.

Less than 2 per cent of the continent's landmass is free of ice. This includes nunataks (boulders of bare rock that stick out from the ice), stretches of coastline on the Antarctic Peninsula, and a series of west-to-east-running basins on the coast opposite Ross Island. Scott's western party had discovered the Taylor Valley, one of these basins and one of three major valleys that make up the McMurdo Dry Valleys, the continent's largest continuous expanse of exposed arid desert soil.

Some of Shackleton's men visited the valley a few years later but it wasn't until February 1911, during Scott's final expedition, that it was explored in detail by Griffith Taylor – and subsequently named after him. The following year, Scott's northern party narrowly missed discovering the neighbouring valley. They sledged along a glacier but never climbed high enough to get a view across from its crest.

That discovery would have to wait for almost half a century and a small band of young Victoria University scientists and students from Wellington. On December 10, 1958, British-born physicist Colin Bull, biologist Richard Barwick, and undergraduate geology students Barrie McKelvey and Peter Webb (back for another summer on the ice) hitched a ride with a helicopter from the fledgling US programme at McMurdo Station. They were dropped off in the as yet nameless valley next to the Taylor Valley and left to their own devices for the rest of the summer. All they had was an aerial image taken during one of Admiral Byrd's reconnaissance flights in 1947, crude maps of the coastline drawn up by Scott's and Shackleton's parties, and boxes full of food and equipment they had begged and borrowed from New Zealand companies. This was the first university-sponsored science party, and Bull had pulled off the venture with little money but a lot of audacity.

The quartet were the first to walk across the valley and to map substantial parts of it. Bull named the valley after Charles Wright, a Canadian scientist on Scott's last expedition, and the glistening lake in its middle after Vanda, the lead dog on one of his own expeditions to Greenland in 1952. What's more, if Bull had had his way, Antarctica would also feature a Mount Cadbury and several other peaks with corporate names. The New Zealand Geographic Board declined his request and instead allowed the team to name just one mountain Sponsors' Peak.

Sculpted rock gardens

Nearly a century after Scott's discovery, my first glimpse of the Dry Valleys was from the air. For a few moments, as the helicopter was turning inland towards the Wright Valley, I could see the entire landscape. The Transantarctic Mountains

The view of the Wright Valley, towards Lake Vanda, from Bull Pass. ROB McPHAIL

block Antarctica's giant ice sheet from flowing towards the coast and the only glaciers entering the valleys are smaller alpine ice tongues, gleaming almost painfully white and surprisingly free of any rock debris. They dominate the picture, with some halting abruptly at the head of a valley and others flowing more gently down the sides. For millions of years, this area has seen next to no rain. It snows occasionally but the air is so dry that most snowflakes evaporate before they reach the soil. They sublime, melting directly into vapour instead of liquid water.

Over the years, geologists have proffered different explanations for how the Dry Valleys have been created. Early accounts, going back to Scott's science team, credited the glaciers as the main force that carved out the basins. This view prevailed until after the International Geophysical Year into the early 1960s, but more recent interpretations suggest that the valleys may have been shaped at least partly by ancient rivers, and that glaciation modified an essentially fluvial landscape.

Our helicopter landed on Bull Pass, named in honour of Colin Bull, which straddles the Olympus Range and connects the Wright and Victoria valleys.

Scientists are dwarfed by
a wind-carved boulder at
Bull Pass. ROB McPHAIL

Immediately, another landscaping force became obvious. Huge cavernous boulders stand like exhibits in a sculpture garden, hollowed out, chiselled and smoothed by a combination of ferocious salt- and sand-laden winds and relentless cycles of freezing and thawing.

During the winter months, when temperatures drop to minus 40 degrees and below, everything is frozen solid. But during the short summer the dark rocks and pebbles absorb the heat and can get warm even by temperate standards, even though the ambient air temperature may only be a little above freezing. Some rocks can warm up to 25 degrees (77 °F) for short periods, only to drop below zero (32 °F) again during the early morning hours when the sun sits a little lower in the sky. Daily temperature fluctuations of 30 degrees (54 degrees in Fahrenheit) are not unusual.

Even though there is very little moisture, each minuscule droplet of salty water becomes a stone-carving tool as it expands every time it freezes and the salt crystals begin to eat into the rock. Slowly, the rocks weather into extraordinary shapes, set against a backdrop of parched land in the planet's most extreme cold desert.

In a landscape that receives less than 50 millimetres (2 inches) of precipitation per year – always in the form of snow, much of which evaporates on its way down – open water is an unexpected component. Yet each valley is dotted with ice-sealed lakes and ponds and, for a few days or weeks during the height of summer, trickles of glacial melt water combine to form ephemeral streams. In many ways, the landscape seems more familiar than the frozen white vastness that frames the view, but at the same time it remains unfathomable.

One of the most alien aspects is the strange geometrical pattern that covers large tracts of soil in each valley, like a giant grid of irregular ring shapes. The pattern can cover several acres, and the largest of the hexagons or pentagons within it stretch across more than 30 metres (100 feet, 33 yards). These polygons form when permafrost soil slumps and the lines around each ring become small trenches that catch wind-blown sand and sediment. In time, they grow to mounds that can be a metre (3 feet, 1.1 yards) high, adding a three-dimensional character to the landscape.

During my visit in late spring, the valleys' riverbeds were still dry, but from the gentle ridge of Bull Pass I could see the sapphire-blue Lake Vanda, our destination for the day. Like any other time I tried to judge distances in Antarctica, the crisp air again disabled my perception. What looked like a two-hour walk would take us most of the day. As we picked our way through fields of wind-sculpted and broken rocks and each of my steps sank a few millimetres (0.3 inches) into the pebbly frozen soil, I was acutely aware of how few people have walked across this landscape before. Antarctica's human history is short and the number of visitors to the Dry Valleys is still counted in thousands.

I had come to find out about the biology of the valleys, and to see one of its famous mummified seals. It wasn't difficult to find one. There are hundreds of seal carcasses scattered throughout the valleys and beyond, and one specimen lies on a mound of rocks close to the gleaming Lake Vanda. Wind and sand have cleaned its bones which are barely covered with patches of desiccated and bleached skin. Small grains of sand have settled in rounded joints and eye sockets. The seal has probably been in its resting place for hundreds of years.

Some seals have lumbered 30 kilometres (19 miles) to their death, and judging by the thin layers of blubber on carcasses that haven't yet decayed completely, they all starved. Although the dominant species in this part of the Ross Sea coastline is the Weddell seal, most of the carcasses in the valleys are crabeater seals and, perplexingly, juveniles. Why they made the long journey inland remains a puzzle, but physiologist Gerald Kooyman has put together a few of the pieces.

Before he shifted his attention to emperor penguins, he studied seals. He was more interested in the living than the dead, but also curious to understand what

Abraded, pitted, etched, grooved and polished by wind-driven sand and ice crystals, large ventifacts define the landscape at Bull Pass. ROB McPHAIL

was driving some youngsters to embark on their fatal marches. During one of his field seasons back in the 1960s, he found a crabeater seal pup moving inland across the McMurdo Ice Shelf. Whenever he attempted to turn it towards the sea, it attacked and continued south. On another occasion, his team captured a Weddell seal moving south through the hills at Cape Evans and released it facing east. It promptly turned around and headed south again.

Such single-mindedness suggests an imprinting process, similar to that first discovered in geese by the Austrian Nobel laureate Konrad Lorenz during the 1960s. Lorenz used incubator-hatched geese to show that they follow the first moving object they encounter within a few hours of hatching, regardless of consequences. Biologists have speculated that a similarly inflexible process, combined with a young seal's lack of experience, could be responsible for their stubbornness. In the case of crabeaters, which are thought to migrate inshore from their birthplace on the pack ice, Kooyman also proposed that some are simply trapped when the ocean begins to freeze again and they head in the wrong direction in search of water. Like whale strandings along temperate coastlines, these misguided seal migrations in Antarctica continue to exercise the imagination of zoologists. From a microbiologist's perspective, however, they provide welcome natural laboratories that have helped to yield surprising insights about life on a microscopic scale.

Age of bacteria

When Craig Cary landed in Antarctica, bound for his first field trip in January 2002, he was more familiar with microbial communities living in extremely hot and toxic environments. His research at the University of Delaware had focused on the microscopic single-celled organisms found around deep-sea volcanoes and their hydrothermal vents, which spew out super-hot, mineral-rich water and plumes of noxious gases. He knew that temperatures of 120 degrees Celsius (248 °F) were no impediment to bacteria colonising the vent systems, and was curious to see what he would find at the other end of the temperature spectrum.

During a sabbatical with Waikato University's thermophile research unit, Cary joined the team for a summer field season in the Dry Valleys. In his luggage was a portable genetic analysis kit that would allow him to check the soil for different types of bacteria with a common technique known as Polymerase Chain Reaction, or PCR, which amplifies any fragments of DNA found in the sample. His kit included 16 genetic tags, or primers, with each of these short bits of genetic code representing the molecular equivalent of a signature, unique to a specific type of bacteria.

Cary remembers setting up his PCR machine inside a polar dome tent and thinking that he would be lucky if the soil samples included a matching piece of DNA for one or two of the bacterial signatures. When all 16 returned positive results, his first thought was that he must have contaminated his equipment.

Molecular genetics had only just arrived in Antarctica. Before the advent of DNA fingerprinting, microbiology relied on culture methods and microscopy, and so far, most bacteria that had been identified in the continent's mineral soils were those that microbiologists managed to keep alive in test tubes filled with a nutritious broth. The only way to identify microbes was to inoculate a sterile culture broth with a drop from the sample and to wait for any cells to divide often enough to show up as colonies in a Petri dish or clouds of life in a test tube. Even if the broth recipe resembled the mix of nutrients in the soil sample or had been mixed to test certain metabolic properties of the organism, a bias towards more common generalists was unavoidable.

Craig Cary prepares to test soil samples in search of microbes. ANTARCTICA NZ PICTORIAL COLLECTION: CRAIG CARY/K023 07/08

This approach yielded a few cosmopolitan groups of soil bacteria and, initially, microbiologists thought that such low diversity was a true reflection of microbial life in the cold desert. It seemed plausible that an extreme environment could support only a few organisms that had evolved coping strategies to deal with temperature fluctuations, extreme desiccation, low nutrient levels and high salinity. As it was difficult to detect any active metabolism in the soil, the assumption was that there was indeed little, if any, life. Descriptions of Dry Valley soils as 'almost sterile' persisted in the scientific literature into the 1970s.

By that time, microbiologists worldwide were beginning to revise their ideas about the ubiquity of bacteria. The earliest molecular techniques were showing that these simple organisms lived essentially in every spot that could support life, including the deep ocean and acidic hot springs. Cary had been part of this inventory update of microbial life during his postgraduate studies at the Scripps Institution of Oceanography during the early 1980s. It was not long after submarine volcanic systems had been discovered and the exploration of bacterial life around these scalding underwater vents had taken off. This period also coincided with the dawn of molecular genetics, which was revealing thriving bacterial populations in places that had been considered off limits to life.

Craig Cary and team relaxing at their campsite in the Miers Valley. Cary is in the middle and Don Cowan second from right. ANTARCTICA NZ PICTORIAL COLLECTION: CRAIG CARY/K023 05/06

Now, similar revelations were emerging from the frozen continent. Cary's PCR machine returned the same results again and again, confirming that the microbial diversity in the Dry Valleys was much higher than had been thought – in fact, unprecedentedly high for an extreme environment – and that previous descriptions reflected the limits of culturing methods rather than the limits of life itself. He teamed up with Don Cowan and Stephanie Burton, who are both based in Cape Town, South Africa, and had begun monitoring microbial biomass in Dry Valley soils by measuring levels of a molecule called adenosine triphosphate, or ATP. ATP is ubiquitous, found in every cell from the simplest bacteria to humans. It is a molecular power plant, storing and transferring energy during chemical reactions within a cell. In the process, ATP is constantly recycled and its levels remain steady, which makes it a useful proxy to measure the number of living cells in a sample.

Cowan and Burton surprised themselves when their experiments showed 10,000 times more cellular life than had been predicted for Antarctic soils. While the early culture studies found perhaps a thousand bacterial cells per gram (0.03 ounces) of soil, their experiments suggested that the numbers were closer to a million, possibly up to a hundred million, cells per gram (0.03 ounces).

Compared to agricultural soils from temperate regions, which usually accommodate up to a billion microbes per gram (0.03 ounces), biomass in Antarctica's soils remains one or two orders of magnitude lower, but nevertheless much higher than previous estimates.

The next task was to draw up a who's who of Antarctic soil microbes – and that's where Cary's PCR signature files confirmed that there were many more different types than those that grow happily in a Petri dish. The team's work confirmed the presence of what microbiologists had come to call the uncultured group.

Cary's first season on ice forged an ongoing interest and he now divides his time between field trips to microbiological habitats at either end of the survivable temperature scale. While he continues his exploration of venting submarine volcanoes, he has returned to the Dry Valleys each summer to figure out what drives microscopic life in the frozen soil. One summer, he began to look more closely under a mummified seal.

Secluded hotspots

In 2006, Craig Cary's team became the first to gain permits to move a seal carcass as part of a science experiment to monitor microbes in the soil. The parched animal they picked had died young and had been in its location for some 250 years, slowly trickling carbon, nitrogen and other nutrients into the soil below. Unlike in temperate regions, where bacteria quickly break down any dead tissue into a stinking, liquefying mass, Antarctica's dry cold air sucks out most of the water first, freeze-drying the carcass. It decomposes slowly and becomes a long-lasting source of nutrients to the soil. The microbial community Cary found underneath the seal was vastly different from the mix of species found in the open soil just yards away. Microbes that are barely detectable in the surrounding soil had flourished to dominate the habitat, changing the composition radically.

Sealed in full bodysuits and surgical gloves and doused in ethanol to avoid contamination, Cary's team lifted the seal and shifted it to a pristine site – and waited. Each year, the team returned to analyse the soil underneath and around the pile of bones and tissue, but they didn't expect anything to happen quickly. In Antarctica, the rate of carbon turnover was thought to be excruciatingly slow, measured on a scale of hundreds of years or even millennia.

Life as we know it is based on carbon. The element provides the chemical backbone for all organic molecules within a living cell, from simple sugars to the doubly twisted strands of DNA. Bacteria and fungi are the main agents in the recycling of

Craig Cary's team was the first to gain permits to shift a mummified seal and to describe the microbial life underneath the carcass. The seals (right and on following pages) have been in their resting places for centuries, slowly mulching the soil with carbon and other nutrients. The seal carcasses also keep the soil moist and protected from the wind. ANTARCTICA NZ PICTORIAL COLLECTION: CHARLES LEE/K023 07/08

carbon from dead tissue back into food for the living, but on ice, this process was thought to be close to standing still. Cary should not have expected to witness any change in his lifetime, let alone in the few years he was planning to dedicate to this experiment. What he found, however, was that the microbes began adjusting to the new, food-rich conditions almost immediately. After only two years, the mix of microbes had changed remarkably, almost resembling the web of microscopic life found in the original site.

Again, the discovery was only possible because genetic techniques have become sensitive enough to detect bacteria that don't grow easily in a test tube and may only exist in very low numbers during dormant stages, when they are completely inactive. Cary's interpretation of his results is that the microbes that become dominant under the seal are always present in the dry soil, surviving in a state of suspended animation, waiting for conditions to improve.

When a seal carcass lands on top of them, food suddenly becomes plentiful and they bounce back to life quickly. While food is a prerequisite in the depleted soils of the Dry Valleys, the seal also provides additional protection. It blankets the frozen soil and allows moisture from the permafrost layer about 30 centimetres

The mummified seal on the left was found near New Harbour; the other on the ice at Lake Bonney in the Taylor Valley. EMILY STONE/ NSF; CHRIS KANNEN/NSF

(12 inches) below to seep up to the surface. Without the cover, water evaporates as quickly as it moves up, but under the seal the relative humidity can change from 30 to 80 per cent, making life vastly more comfortable.

The seals provide a model system for microbial degraders – bacteria that depend on an external food source. But they are mere hotspots of carbon, like patches of tropical forest in the middle of a desert, in an otherwise nutrient-poor environment. Dry Valley soils are the result of weathering of bedrock and glacial tills. They are extremely dry, equivalent to some of the world's hottest deserts, but they are saltier than temperate soils because there is no rain to flush the mineral out. However, in contrast to hot deserts with extremely low rainfall, water is always close. Massive slabs of buried ice, cemented with the soil, run underneath about half of the area in the Dry Valleys. As mentioned above, this permafrost layer is usually about 30 centimetres (12 inches) deep, and somewhere between this layer and the surface the soil can be relatively humid – at least moist enough to offer a tantalising possibility for life to gain a foothold. The highest concentrations of living cells and organic matter are found at about 5 to 10 centimetres (2 to 4 inches) below the desiccated surface, but even here the nutrient levels are extremely low and microbial communities depend on a do-it-yourself strategy, known as autotrophy, to get started.

Like plants, some bacteria can use the sun's light to make their own food by sequestering carbon from the air. Others draw their energy from chemical reactions to fix either carbon or nitrogen. Collectively, they are the founding members and primary producers, the first link in the microbial food web, because they convert inorganic carbon and nitrogen into food for the degraders that cannot make their own.

Beyond these first levels of the food web, terrestrial microscopic life in Antarctic soils appears to be simple. Although biological diversity – the number of different types of microbes – is higher than previously thought, biomass, or the actual number of living cells, is low. In contrast with less extreme habitats, where food webs usually have several layers of hierarchy all woven together in a complex system of predator-prey or competitor relationships, microbial life in the dry Antarctic desert is ruled largely by non-biological factors. At its simplest level, the community is exclusively bacterial and represented by a few primary producers and degraders. They live together rather than off each other, and depend on at least occasional availability of liquid water. Anything more complex is rare and limited to the more moist soils near streams and lakes or microscopic habitats that are more protected from the elements.

One such micro-niche is under a rock. Depending on the geology, some areas of the Dry Valleys are littered with small quartz or marble pebbles and, typically,

Measuring the flow of
carbon dioxide underneath
a mummified seal.
ANTARCTICA NZ PICTORIAL
COLLECTION: CRAIG CARY/K023
07/08

underneath any stone that is bigger than a quarter, microbes thrive. The stones are translucent enough for light to penetrate but dark enough to hold a little more moisture underneath, providing much more pleasant conditions than the loose grains.

Similar microbial communities have colonised cracks and fissures within stones, and even the tiniest spaces inside coarse crystalline rock types, such as the Beacon sandstone layers that are typical of the northern Dry Valleys. Inside these rocks, protected from the wind and buffered from fluctuating temperatures, bacteria often mingle with fungi, algae and lichens.

Like the patches of soil under seal carcasses, these rocky niches are an anomaly in the Dry Valleys. The most common – and harshest – environment is the open soil. Perhaps not surprisingly, the microbes Cary's team is finding in the latter are different from any other habitat, even different from those found in soils sampled in other parts of Antarctica. At the highest taxonomic level, many belong to common groups and are related to desiccation-resistant, cold- and salt-tolerant microbes that can be cultured, but under closer scrutiny, at a species level, many have never been seen before.

Ancient life

Time permeates the Dry Valleys on different scales, like an ever-present fourth dimension. The mountains are the repositories of deep time, reckoned in hundreds of millions of years. High above the valleys, the manila Beacon sandstones are a legacy of Antarctica's warm past, while further down the slope, dark intrusions of fire-formed Ferrar dolerites are signatures of the upwelling of molten rock from beneath the Earth's crust some 170 million years ago. On the valley floors, five-million-year-old volcanic deposits dominate.

In stark contrast, the permanently ice-covered lakes and ponds that are dotted throughout the landscape are thousands, perhaps only hundreds, of years old. The melt streams that feed them sometimes last only for a few days or hours. Yet it is in these most ephemeral habitats that biologists have found the most ancient forms of life.

Cyanobacteria reach back more than three billion years to a young Earth and the very beginning of complex life. Like all other bacteria and another ancient group of single-celled organisms called archaea, cyanobacteria are prokaryotes, the simplest and smallest living things on Earth. They are not much more than a tiny membrane-sealed bag of molecules. Without a nucleus, their genetic code is stored in a simple loop of DNA that floats freely inside the cell.

Yet, however simple, these tiny organisms were major players in the drama of Earth's early evolution. They were the first to master photosynthesis – drawing on the sun's energy to combine water and carbon dioxide into large sugar molecules. Oxygen is a waste product of this process, and so, almost as a side effect, cyanobacteria continued to exhale the gas into Earth's early atmosphere to slowly change it to the air we breathe. Without them, no oxygen-breathing creature could have evolved and we simply wouldn't be here today.

Cyanobacteria are also at the centre of a theory known as endosymbiosis, which explains how more complex life developed by swallowing prokaryotic organisms and using them as power plants to fuel the larger structure. Ancient photosynthesising cyanobacteria are thought to have lent their services to the earliest proto plant cells and gradually evolved into what we now know as chloroplasts.

Three decades after his first field trip to the Dry Valleys, Clive Howard-Williams still remembers watching a desiccated frozen mat of cyanobacteria spring back to life in a drop of water. It was 1982, early spring, and he and his colleague Warwick Vincent had arrived in the Wright Valley to study the short-lived but powerful Onyx River. Elsewhere, the Onyx might be no more than an obscure stream, but in ice-wrapped Antarctica the 30-kilometre (19-mile) river qualifies as the longest stretch of flowing water. But it flows only for a few weeks during the southern summer. For the rest of the year, its bed is dry and frozen, and the idea of a burbling stream seems most improbable.

Howard-Williams' research interest is in freshwater habitats, but his initial focus had been on the tropics, including wetlands in east Africa and floodplains in the central Amazon. He arrived in New Zealand during the late 1970s to join the freshwater ecology group at what was then the Department of Scientific and Industrial Research (DSIR) in Taupo and to study pollution and algal blooms in the Rotorua lakes. But the DSIR also had an Antarctic programme and needed freshwater ecologists to study nutrient cycles in streams and rivers on the icy continent.

Howard-Williams didn't hesitate – and promptly found himself in a landscape he had no eye for. Scanning the dry floor of the Wright Valley, he found it difficult at first to believe that he was standing in a streambed. As he was walking along the main river channel and its tributaries, all he could see was a desert landscape of randomly strewn boulders. But then he began to spot small flat fragments that looked like bits of frozen cow dung, dried up and curling at the edges, covering some pebbles and crevices. He recognised them as latent freeze-dried microbial mats of cyanobacteria, held together by a sugar-based glue which they excrete.

This early in the season, after months of complete darkness and with temperatures still at minus 15 degrees Celsius (5 °F), the mats remained in wintering mode and showed no signs of life. But a few drops of water were enough to resuscitate the thin film of bacterial cells. Within ten minutes they were starting to take up carbon dioxide, and within 24 hours they had bounced back to a fully functioning, photosynthesising freshwater stream community.

Howard-Williams and Vincent moved field camps to the Taylor and Miers valleys to peg out other sites of interest, and they visited the penguin colonies at Cape Bird to study streams flowing close to a rich source of nutrients. When they returned to the Wright Valley a few weeks later after the Onyx River had started flowing, the entire place had burst back to life, much like a hot desert erupts into flower after rainfall. Here, though, the blooming was on a microscopic scale. The once leathery mats had rehydrated into spongy carpets of active cyanobacteria – and a simple but effective way of sampling was to cut out standard circles with a cut-off syringe.

With the help of a sealed chamber, Howard-Williams set about measuring the mats' physiological activity. He monitored their uptake of carbon, nitrogen and other nutrients and found that they were acting as highly efficient filters, stripping nutrients from the glacial melt water as it flows down the Onyx through small sandy braids and across a vast, flat pavement of boulders that provides perfect habitat for the bacteria. By the time the Onyx flows into Lake Vanda, the water is on a par with the purest in the world.

Lake evolution

The Onyx River is the single big inflow into Lake Vanda, but the lake has no outflow. Instead, the lake's levels rise and fall with the often spectacular fluctuations of the river's flow during the summer season.

In his office at NIWA in Christchurch, Clive Howard-Williams still keeps bulging folders of river-flow data collected since the summer of 1969/70 when the then Ministry of Works set up flow gauges and weirs along the Onyx as part of a nation-wide river monitoring programme which included the Ross Sea Dependency. The data set has changed hands since – to NIWA at first and then to the US team involved in long-term ecological monitoring in the Dry Valleys – but the measurements have continued throughout and built up to the longest flow record of any Antarctic river.

The graphs show that, in an average year, a trickle begins to flow at the start of summer, builds to a peak by January, and peters out quickly a month or so later. The Onyx can swell to a sizable river, with average flows of 200 litres (53 gallons) per second, but the amount of water fluctuates daily depending on how clear the sky is and how much sunshine reaches the glacier surface. Annual changes are even more dramatic, varying from completely dry summers to flood years that overflow the main channel and can overwhelm the gauges. Naturally, as Howard-Williams was following the Onyx River's flow into Lake Vanda, his attention soon turned to the lake itself, and the bacterial mats within it.

Lake Vanda today is a small version of a once enormous body of water. With no outflow, any water loss is governed by sublimation of ice or evaporation of melt water during summer. The lake's bottom water is saltier than the Dead Sea – and it is these salts that have recorded the lake's long history. Geochemists managed to date the salts and found that Greater Lake Vanda may have once filled the entire valley, until it began to evaporate after the last glaciation. It continued shrinking until, some 1500 years ago, it was no more than a shallow, highly concentrated

Hydrologists measuring the flow of a glacier tributary of the Onyx River. ANTARCTICA NZ PICTORIAL COLLECTION: NIGEL S. ROBERTS/SCH65

brine pool. When it began to fill up again, a little over a thousand years ago, the dense and heavy salty water stayed at the bottom.

Today, Lake Vanda is about 5 kilometres (3 miles) long and 75 metres (250 feet) deep, and permanently covered with 3 to 4 metres (10 to 13 feet) of the smoothest and most transparent ice I have ever seen. Its water is strongly stratified into a series of density layers, suggesting that the lake has gone through several phases of evaporation and refilling, with each layer of salty water reflecting historical rises and falls. The fluctuations in water level continue to this day, reshaping the landscape so dramatically that a research station built by New Zealand in 1967, Vanda Station, had to be pulled out again less than three decades later, partly because the lake had risen too close to the huts.

During December, the lake's shoreline melts to form a moat of pure potable water, but for the biggest part of the year the entire lake is under ice. In contrast to lakes in more temperate areas whose surface is usually the warmest, Lake Vanda is a giant solar heater, trapping the warmest water at the bottom. Warmer water would normally rise to the top but the noxious brine at the bottom of Lake Vanda is

Brilliant blue Lake Vanda is one of several permanently ice-covered lakes found throughout the McMurdo Dry Valleys. ROB McPHAIL

Clive Howard-Williams
demonstrates the depth
of Lake Vanda's ice cap.
ANTARCTICA NZ PICTORIAL
COLLECTION: TIM HIGHAM/
SCH128

A geology camp near
Lake Vanda, Wright Valley.
ROB McPHAIL

more saline, and therefore denser and heavier, than the other, fresher layers. Any heat from sunlight that filters through the lake's icy lid and down to the bottom is captured, inescapably. At 24 degrees Celsius (75 °F), the lake floor is almost hot by Antarctic standards.

Howard-Williams and his team found that photosynthesising micro-algae and cyanobacteria dominate carbon fixation in the lake as well, to a considerable depth. The extraordinary clarity of the water, indistinguishable from that of distilled water, allows light to penetrate to great depths in the lake, and despite the thick ice cover the peak in plankton biomass and photosynthesis occurs at 63 metres (207 feet). This deep layer of algae and cyanobacteria is perched just above the point at which the lake water rapidly goes from having plenty of oxygen to having none. To scientists this is known as an oxycline, and, just below this boundary, bacteria that decompose and recycle inorganic material thrive. The phytoplankton hovers right above, waiting for the nutrients to diffuse up into the illuminated part of the lake where they re-fix them into organic material.

Rubbery mats of cyanobacteria cling to the rocks as trickles of water begin to flow during the Antarctic summer.

ANTARCTICA NZ PICTORIAL COLLECTION: PHIL NOVIS/ K124 03/04

This deep layer of microbes is, in many ways, the lake's protection squad, as Jenny Webster-Brown found after tracking the fate of trace elements such as iron, copper, zinc, manganese and arsenic along the water column. The former DSIR geochemist, now at the University of Canterbury, has worked in tandem with both Vincent and Howard-Williams, and her initial interest was to establish what happens to a lake's chemistry during the dramatic changes in volume as it goes from a large body of pure fresh water to a brine that can be several times saltier than sea water. She was curious to find out if Lake Vanda was typical of other stratified Antarctic lakes and could be used as a model of lake geochemical evolution.

At the time of her first field season in 1987, Vanda Station was still being used and there was concern about the possible impact on the lake of contamination from the research facilities. Grey Water Gully, for example, is named for the function it had during the station's early years, before the introduction of an environmental code of conduct which now prevents any disposal of waste, however harmless, anywhere. When the station was removed, Webster-Brown had a chance to test the soil and water around the site for any trace metals. She found they were elevated at the site but, with the exception of small amounts of zinc, were too tightly bound to the soil to move into the lake water. If Lake Vanda had been contaminated, she expects it would have coped better than other more temperate lakes because of its permanent oxygen-depleted bottom layers and the sulphur bacteria that thrive in these toxic brines. As part of their metabolism, these microbes help to produce insoluble metal-sulphide minerals, and act as effective clean-up agents.

Recently, Webster-Brown has shifted her attention to the geochemistry of smaller freshwater habitats, ponds and cryoconites, which may last for only a few seasons. She remembers flying over glaciers, watching hundreds of tiny brilliant blue ponds and wondering whether they were all the same chemically. Countless such pools form when the sun warms up specks of soil and rocks, which then melt downward into the surrounding ice. The resulting water-filled hollow eventually freezes over again, creating a miniature aquarium of microbes that had been lying on the surface, dormant but ready to bounce back to life in water. Collectively, cryoconites may well represent the largest volume of liquid water in many catchments and act as refugia for freshwater communities.

Ponds are generally larger than cryoconites and can appear on land or on ice. Usually, ponds don't have a distinct inflow but are close to a snow bank or form when melting permafrost pools in a small basin. On glaciers, some ponds are the result of several cryoconites combining into one larger puddle.

The construction of the first hut at Vanda Station in October 1967. The first Victoria University expeditions during the International Geophysical Year had triggered international interest in the Dry Valleys; American, Japanese, Italian and Russian teams were all working in the area and Vanda Station was built to allow teams to winter over. By the time Clive Howard-Williams landed in the Wright Valley in 1982, new solar-powered huts were about to be added to the station and the original buildings upgraded – except the toilet hut which had no door but a fabulous view. The station also hosted the Royal Vanda Swimming Club, begun by Colin Bull, which played an important role in station morale over the years. Friendly rivalry existed between the seasonal staff (Vandals) and the Ministry of Works hydrology/glaciology team (Asgaard Rangers) who worked from the station. However, by the early 1990s the lake had risen too close and all buildings were decommissioned. By 1995, all traces of the station had been removed, but analysis of the lake water and algae continued to ensure that the lake had not been contaminated. ANTARCTICA NZ PICTORIAL COLLECTION: WAYNE ORCHISTON/BV3

Some are as fresh and pure as newly melted ice, while others on land have evaporated into a strong brine. It can be easy to tell which is which by a significantly thinner ice cap and a ring of white salt crystals around the edges of more saline ponds. The salts are a chemical signature of the local environment. Near the Darwin Glacier further south, Webster-Brown found ponds surrounded by a rare nitrogen-rich mineral called nitratine, reflecting the influence of the upper atmosphere on pond chemistry in Antarctica's interior. She also found ponds with surprisingly high concentrations of trace metals such as arsenic and uranium, derived from basement rocks in the area.

One region that is regularly perforated by interconnected ponds is Bratina Island in McMurdo Sound. This island sits near the centre of the moraine-covered McMurdo Ice Shelf, an area known as the 'dirty ice'. This part of the ice shelf has been described as Antarctica's largest wetland because the surface melts out for a few months each year to form thousands of pools and small streams, many teeming with mat-forming cyanobacteria. Howard-Williams suggests that the McMurdo Ice Shelf is a seeding ground for freshwater life in the region, courtesy of katabatic winds that pick up freeze-dried microbial mats during winter and transport them along the coastline and inland of Victoria Land.

Lakes without fish

The DSIR group was one of the first to work on the microbial and algal physiology and ecology of Dry Valley streams and lakes, but on the other side of the continent, on an island near the coast of the Antarctic Peninsula, Ian Hawes was exploring similar habitats for the British Antarctic Survey.

At the end of his studies at the University of Liverpool, Hawes had been left with two options: tropical gastropods in the Seychelles or microbial mats in Antarctic lakes. He opted for the cold and, in October 1978, at the age of 21, sailed south to spend two and a half years on ice. He knew almost instantly that he had made the right decision. The experience of the continent's dramatic seasonal changes and being able to watch as life hunkers down for the long polar night has filled him with a deep affection for Antarctica that has never faded.

Hawes' research focus was on nutrient flows in freshwater ecosystems. The Antarctic Peninsula is climatically distinct from the rest of the frozen continent, and its lakes support more complex communities and a larger slice of the food web. Small freshwater crustaceans are common and in some ways the peninsula's lakes resemble temperate ones, except that they have no fish. In Hawes' study area, seals

were staging a comeback from hunting and adding nutrients to the catchment, fertilising the lakes. Lake eutrophication (nutrient enrichment) had been recognised as a major environmental issue worldwide and he was curious to see how colder and biologically less complex habitats would cope.

He returned to Britain to complete his PhD, which led to another winter in Antarctica and shifting his focus to the microbes that survive in the murky sediment at the bottom of lakes. Not long after a visit to New Zealand to present his research at a conference, Hawes received an early-morning phone call and a job offer from Clive Howard-Williams (then with NIWA's predecessor, DSIR), which was the start of an ongoing collaboration and a new quest to fathom how life works in the even more isolated lakes in an Antarctic oasis.

Hawes is a diver. Instead of sampling lakes from the surface, he explores them close up. He has spent hundreds of hours under water, many of them somewhere in Antarctica, but the permanently ice-covered lakes have provided the most other-worldly experience, even in comparison with his countless explorations below the frozen ocean. For a start, Antarctic lake divers generally operate alone,

ABOVE AND OPPOSITE
The complex requirements
for diving in Lake Hoare.
The dive mask seals the
entire face, but allows the
diver to communicate
with the support team on
the surface via a cable.
ANTARCTICA NZ PICTORIAL
COLLECTION: HILKE GILES/K081C
07/08.

tethered to the surface for safety. Solo diving is one of the measures scientists have taken to protect the ecosystem from too much disturbance, particularly in lakes whose water column is stratified and each layer of increasingly salty water is likely to support a distinct mix of organisms. Hawes doesn't mind. Without a dive buddy to worry about, he can relax and focus his attention on the extraordinary environment.

Most Antarctic lake dive teams have also abandoned scuba tanks in favour of unlimited air supply via a hose from the surface. In Lake Vanda, Hawes went a step further and used a rebreathing system which recycles exhaled air to prevent bubble plumes from mixing up the water layers and thereby minimises any environmental disturbance.

The crystal-clear Lake Vanda may be one of the most comprehensively studied bodies of water in Antarctica, but its smooth, hard and perfectly transparent ice

surface is rare. Much more light can penetrate its ice cap to create a distinct ecology. In contrast, Lake Bonney, at the western end of the Taylor Valley, is known for long, columnar gas bubbles that are encased in its ice. Lake Hoare, damned by a tongue of the Canada Glacier, is covered with rough patches of sand. And further along the Taylor Valley, Lake Fryxell is stratified like Lake Vanda but its frozen surface is covered with sand and, around the fringe, with mounds of cyanobacteria that have detached from the lake sediment and floated up, slowly working their way through 6 metres (19 feet, 6.6 yards) of ice.

However, all Dry Valley lakes share some common features. Like desert lakes in hot regions, they are usually closed systems with no outflow. Their icy lids eliminate wind-generated currents and any mixing of the water column, reducing light penetration and restricting gas exchanges between the water and the atmosphere. Of particular interest to Hawes was that the lakes harbour photosynthesising

Lowering instruments to
the bottom of Lake Hoare.
ANTARCTICA NZ PICTORIAL
COLLECTION: HILKE GILES/K081C
07/08

bacterial communities, often dominated by cyanobacteria, which cover much of
the lake floor.

He has explored many of the lakes in the Dry Valleys, but the blue hues of Lake
Vanda have left a lasting imprint. Floating in what he describes as a blue stained-
glass cathedral, he dropped down to 40 metres (130 feet), with crystal water clarity
allowing clear views of the flat underside of the ice above and the spongy mats on
the dark lake bed a further 30 metres (100 feet) below. The silence was absolute.

In contrast, Lake Hoare is gloomy, enclosed, and its plumes of bacteria in the
water column evoke images of cosmic nebulae. But it was in the near darkness
of this lake that Hawes made some astonishing discoveries about the capacity of
bottom-dwelling bacterial mats to fix carbon by extracting energy from the slight-
est residue of sunlight.

When he started his research project, others had already worked on the lake but their interest had focused on floating planktonic microbial communities. Other divers had looked at the bottom-dwellers and published some descriptions, species lists and even photographs, providing evidence that microbes were alive and well in the mud at the lake bottom. However, attempts to measure photosynthetic processes and to show that this was an active community rather than a product of sedimentation had largely failed or sparked controversy. Hawes and his team developed a new technique that allowed them to sample mats on the lake floor without changing their immediate environment.

One of the unusual features of all Dry Valley lakes is that they are supersaturated with various gases, including oxygen. When water freezes, it excludes any gases that may have been dissolved in it and their concentration increases in the water column below the ice. This has made it difficult to measure accurate rates of photosynthesis, which, historically, has been charted as minute changes in oxygen concentrations. Against the high background reading in the lake water, such small fluctuations can become impossible to discern. At the lake bottom the difficulties to get accurate results are multiplied because more gas dissolves under pressure, and it simply bubbles out when a sample is moved up to the surface and pressure is released.

Hawes' solution was simple but effective. He had come up with a sampling device that seals each sample under pressure and stops any gases from escaping. As a result, his team became the first to prove and quantify active photosynthesis in microbial mats on the lake floor, and to show that these organisms had a carbon-fixing capacity that rivalled or even exceeded that of planktonic communities. The results changed the way people thought about primary production in Antarctic lakes. Up to this point, floating microscopic organisms had taken all credit for photosynthetic activity but it was now clear that at least half of it could be attributed to the benthos, or bottom-dwelling communities.

A few years later, I met Hawes on the surface of Lake Fryxell, just as he was emerging from the narrow dive hole after another solo descent. He and his team had spent the first days of their field season setting up camp and melting a hole through 6 metres (19 feet, 6.6 yards) of ice with a water-heated copper coil. But now the hard work was done. Ahead of them lay a few more days of pure exploration. By that stage, the need to bring up samples from the lake bottom had been eliminated completely by a new sampling device the team christened 'Hal'. In contrast to its moody, malfunctioning namesake from Arthur C. Clarke's *Space Odyssey* saga, this automated contraption is a precision tool. It carries several micro-electrodes that can measure minute concentration changes in oxygen or other dissolved gases.

ABOVE Ian Hawes examines
a sample of cyanobacteria,
also known as blue-green
algae, which form mats in
visible layers. ANTARCTICA NZ
PICTORIAL COLLECTION: K081C
07/08

OPPOSITE Bubbles of gas
and debris trapped in the
icy surface of Lake Hoare.
ANTARCTICA NZ PICTORIAL
COLLECTION: HILKE GILES/
K081C 07/08

It can simply be placed on the lake bottom and left there to manipulate the electrodes, checking microbial mats at varying depths and precise time intervals. The result is a continuous profile of the mats' metabolic activity.

The surface of Lake Fryxell is covered with tall, crumbling towers of rough, opaque ice that can be several decades old. Underneath, the water is even darker than in Lake Hoare. Only 1 or 2 per cent of the sunlight at the surface penetrates the ice and it can take a diver several minutes to adjust to the extremely dim light. But even here, bottom-dwelling microbial communities make the best use of the little they get. Their composition changes with depth, from flat mats that are largely made up of cyanobacteria to white and stinking clouds of sulphur bacteria that follow a different photosynthesis recipe. Oxygen is lethal for the latter bacteria, and instead of splitting water to gain electrons (which produces oxygen), they use hydrogen sulphide and turn it into tiny granules of elemental sulphur. These microbes are even better than cyanobacteria at making the best use of very little light.

In the light-limited communities in Dry Valley lakes, the correlation between rates of photosynthesis and the amount of light is direct and linear: the more light filters through the frozen surface, the more photosynthesis can happen. Anything that causes a change in the available light also changes the lake's primary

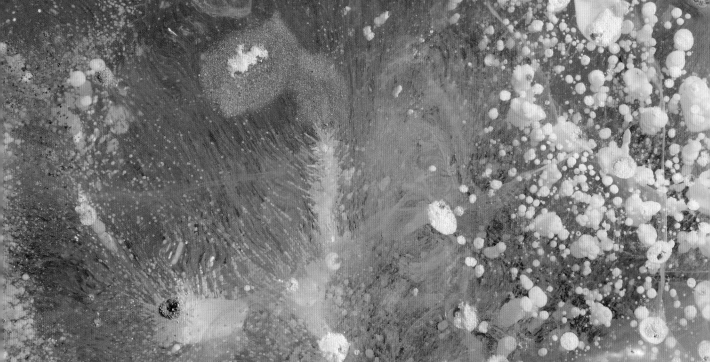

production. A bit more or less snow on the surface can have a significant effect, and Hawes' team has been investigating microbial mats as possible archives of such changes, chronicling past levels of photosynthetic output and, by proxy, recent past climates.

The mats are laminated, building up layer by layer during each short burst of activity. Their thickness is a first approximation of how productive a lake has been in the past, but Hawes has found a chemical marker which tracks past growth spurts even better. The polysaccharides that hold the mats together usually decompose quickly, and as their organic matrix decays, it releases calcium that was bound in the structure. It joins with carbonate and precipitates as calcite, the amount of which leaves behind a clear signature of previous growing conditions.

Life after sunset

John Priscu's introduction to Antarctica began in New Zealand while the Montana State University ecologist was working for the DSIR in Taupo during the early 1980s. He had come to investigate the microbial ecology of lakes throughout the country but, working with Clive Howard-Williams and Warwick Vincent, he soon found himself on board a C-130 Hercules heading for the ice. He had read most of the Antarctic literature, including Scott's diaries and his descriptions of the 'valley of the dead', and still remembers looking through one of the plane's small portholes at the vast icy landscapes below, thinking 'if only Scott had had a micro-scope'. Thriving microbial communities had recently been discovered in the deep ocean and Priscu simply could not believe that the ice should be lifeless.

When he returned to the US at the end of 1985, he yearned to go back to the Dry Valleys and set out to establish a research programme. For a few years, he focused on studying micro-algae that live in sea ice, but by the early 1990s he was back at work on the ice-covered lakes in the Taylor Valley. Like the New Zealand ecologists, Priscu has focused on trying to understand the biology and chemistry of permanently ice-covered lakes on the frozen continent – the only place in the world where such ecosystems form – and how cyanobacteria and other microbes and algae manage to survive in them. The main study site for his team, Lake Bonney near the Taylor Glacier, is fed by a stream named after Priscu.

In 1993, the Dry Valleys were selected as a study site for the US National Science Foundation's Long-Term Ecological Research project, led by Priscu. He soon embarked on a genetic analysis of cyanobacteria mats in the hope of establish-ing how they are distributed in the valleys. The composition of bacteria in the

John Priscu at Bratina Island, with the Royal Society Range in the background. JOHN PRISCU/NSF

mats proved to be similar throughout the area and beyond, and Priscu found that they survive everywhere they can find water, even if it is just for a few days each year. Their habitats range from moist soils to melt streams, lakes and ponds, and cryoconites – the smallest bodies of water, often no more than a tiny ice-covered puddle on the surface of a glacier. This work also confirmed Priscu's hypothesis that wind played a major role in how the mats were being distributed throughout the valleys.

One of his greatest discoveries, however, was to show that microbes thrived not just in liquid water but in solid ice. As Priscu was drilling through Lake Bonney's frozen cap to study the water below, he regularly struck a layer of sediment about 2 metres (6.6 feet, 2.2 yards) from the surface. For many years, it was no more than a nuisance, both in the field and at the campsite (where some of the dirty ice cubes ended up in drinks). Initially, Priscu thought that the sediment had been dumped

Rob VanTreese and Karen
Cozzetto record light level
readings beneath the ice
of Lake Bonney. While
one person dangles the
light-reading instrument at
different depths, the other
team member records
the instrument's reading.
The tarpaulin prevents
sunlight from affecting
the measurements.
CHRIS KANNEN/NSF

on the lake by a particularly violent storm and that it would eventually melt its way through the ice. It never did.

At the time, Priscu was particularly interested in the fate of nitrous oxide, or laughing gas, in Lake Bonney. The lake is loaded with this potent greenhouse gas, with concentrations thousands of times higher than those in air, and he wanted to understand whether it could be released into the atmosphere. Priscu knew that the gas is produced by bacteria, but he expected the microbes to be distributed throughout the lake – until he tested that dark layer of ice. In 1998, his team became the first to describe a thriving community of bacteria living in solid ice.

Despite coming across life in unexpected places, Priscu also established that the lakes themselves are slowly dying. Like Lake Vanda in the neighbouring Wright Valley, the ice-covered lakes of the Taylor Valley are remnants of a much larger body of water that filled almost its entire length about 10,000 years ago. As its

surface froze and sublimed, it left behind highly concentrated brine pools, and much of the carbon and other nutrients that drive microbial life today were deposited then. Priscu's analysis showed that the lakes' microbial communities are still living off that legacy carbon now.

In 2007, Lake Bonney became a testing site for a NASA project with the ultimate goal of sampling the ice-covered oceans of Jupiter's giant moon Europa – one of the few places in our solar system where scientists think life may have had a reasonable chance of establishing a foothold. Each day for a month, the ENDURANCE robot – officially known as the Environmentally Non-Disturbing Underwater Robotic Antarctic Explorer – slipped down through a hole in the ice in the middle of the lake and followed a pre-programmed sampling schedule, recording temperature and taking various chemical and biological measurements. At the end of its shift, the round submersible resurfaced to be plugged in for an overnight battery recharge.

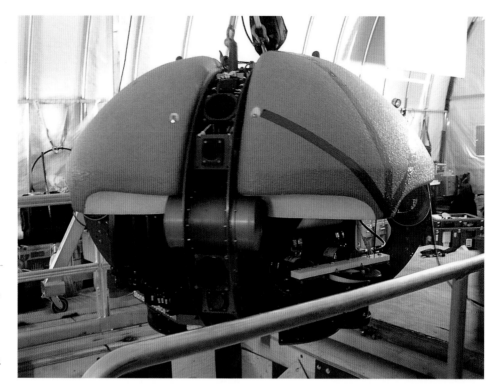

The ENDURANCE robot, which was used to measure temperature, electrical conductivity, ambient light, chlorophyll and other biogeochemical information in Lake Bonney.
LIZ KAUFFMAN/NSF

The NASA project provided a rare opportunity to collect a more comprehensive data set, but Priscu felt that there was still a major gap in the research: the dark period. During the early 1990s, he had twice travelled to the ice on one of the earliest flights in late August, just as the sun starts to appear above the horizon again. His team took samples at minus 50 degrees Celsius (-58 °F) to measure the photosynthetic activity in the lake. The results suggested that the microbial residents remained active during winter, despite total darkness. Ever since, Priscu suspected that the transition at the start of winter, from daylong sunshine to absolute darkness, was likely to reveal new biological processes, and he was keen to stay beyond the usual summer research season. During the International Polar Year in 2007/08, his team and the New Zealand researchers got the first opportunity to monitor ice-covered bodies of water as their main energy source slowly disappeared. The teams stayed beyond the onset of complete darkness and had to be picked up by a C-17 Globemaster whose pilots were flying and landing the plane with night-vision goggles.

Priscu thinks the effort was worth it. He was able to bring back a green alga to Lake Bonney which he had kept in culture in his laboratory for more than two decades. It was returned to the exact spot in the lake where it was taken from during the 1980s, albeit inside a tube that lets light and nutrients through its membrane but remains impenetrable to the algal cells. Priscu had already shown that this alga is the single-celled equivalent of an evergreen tree – it doesn't grow during winter but it remains green. DNA analysis of samples taken from the reintroduced cells showed that it can turn off the genes required for photosynthesis and survives in maintenance mode on stored energy, without losing its pigment.

While the US team focused on lakes, the New Zealanders watched what happens in shallower ponds on Bratina Island. Ponds are usually completely frozen in the depths of winter, but the team found that the freezing and thickening of the icy lid happened gradually. When water freezes it expels salt and other chemicals, and the remaining lens of liquid water turns into an increasingly concentrated brine. Some ponds changed from brackish water to a hyper-saline brine within a few days, yet the microbial mats living in them remained active. Once the sun disappeared below the horizon, the community adapted by changing its composition, shifting from species that need light to photosynthesise to those that can live off nutrients in the water or sediment. In many ways, this transition period is likely the most stressful time for microbes, and freezing may come as a relief.

Antarctica's climate demands flexibility. Like any other terrestrial life forms, microbes have to cope with long periods in a dormant, reversible state of complete shutdown. Many tolerate extreme desiccation and use it as a survival strategy as it removes all water before it can freeze into ice crystals and cause damage.

Further studies have shown that very few species of Antarctic cyanobacteria are truly psychrophilic, or cold-loving, and adapted exclusively to life in frigid temperatures. Almost all photosynthesising bacteria are merely cold-tolerant, which means that their optimal growth conditions are above 15 degrees Celsius (60 °F), but they have evolved strategies to survive and grow at or close to zero (32 °F). Their survival chances are good almost anywhere, supporting the idea of an ancient ecological history. As far as extreme conditions are concerned, cyanobacteria have seen – and survived – it all.

Microscopic ark

Microbial mats that survive in Antarctic lakes today provide a link with life's infant stages. Some of the earliest-known fossils are stromatolites, layered structures thought to have been formed by microbial mats cementing sedimentary grains into their matrix some three billion years ago. They form a significant part of the

Gushing from the tongue of the Taylor Glacier onto the ice-covered surface of West Lake Bonney, Blood Falls provides evidence that microbes can live deep under the ice. About two million years ago, the glacier sealed a small body of marine water beneath it, including an ancient community of microbes. Trapped below 400 metres (1300 feet) of ice, the microbes have remained in this natural time capsule ever since, surviving without light or oxygen. They generate energy from chemical reactions with sulphur and iron, both in abundance in the brine below the glacier. When Australian geologist Griffith Taylor first discovered the frozen waterfall in 1911, he thought the red colour came from algae, but research by John Priscu's team showed that it is due instead to iron oxides and microbial processes.

fossil record for more than two billion years, while life slowly transformed from simple colonies of single-celled organisms to more complex forms. Modern mat-forming cyanobacteria are seen as analogues for these earliest living communities on Earth.

They demonstrate their resilience in the Dry Valleys and in melt pools on the surface of glaciers and ice shelves in Antarctica. They survive in water that is several times more saline than sea water. They cope with total desiccation and freezing, six or seven months of darkness, exposure to high levels of ultraviolet radiation during spring and daily freeze-thaw cycles during summer. Very simi-lar communities thrive in the high Arctic, which led Clive Howard-Williams and Warwick Vincent to argue that such freeze-tolerant mat consortia could indeed be relics from ancient times that have not only survived Earth's major ice ages but also provided shelter for a variety of other, often more complex, organisms within their gelatinous matrix.

They may even help explain how life endured Snowball Earth, a period when much of the planet is thought to have been covered in ice. The theory postulates that Earth's surface froze over completely at least once, most recently some 650 million or more years ago, before the sudden burst in evolution of new life forms known as the Cambrian explosion. It is one explanation offered for geological evi-dence of glaciation in low latitudes, but its critics have argued that life would have been wiped out by freezing on this scale.

Some scientists think cyanobacteria could have survived, and that their mats would have provided refuge for less freeze-tolerant bacteria within a sticky micro-scopic ark. What's more, these ancient communities may have favoured strong chemical and biological interactions between different species packed tightly together – ideal conditions for symbiosis, preparing the ground for the develop-ment of complex organisms through evolutionary time.

Hidden lakes and rivers

In early February 2012, much-awaited news arrived from the coldest place on Earth. A team of Russian engineers and scientists had entered one of the planet's last uncharted ecosystems by drilling through nearly 4 kilometres (2.5 miles) of Antarctic ice to pierce Lake Vostok, a subglacial lake the size of Lake Ontario in the US or Siberia's Lake Baikal. It had taken them nearly two decades and their project has sparked a heated debate about the risk of contaminating a pristine habitat that has been sealed off from the rest of the world for millions of years and pre-dates

The Russian station Vostok, on the Polar Plateau, is the coldest place on Earth. Lake Vostok lies 4 kilometres (2.5 miles) deep under the icy surface. JOSH LANDIS/NSF

the emergence of the earliest hominids. However, it has also triggered the development of new drilling methods, and fuelled international interest in subglacial lakes.

The team began drilling at Vostok Station in 1992. A few years later, when the existence of a giant lake beneath their feet was confirmed, they drilled down to almost 3700 metres (2.3 miles) but halted the project due to international concerns about contamination. When drilling started again, the team's strategy was based on the fact that the lake is under pressure and the expectation that water would rise up the drill hole and freeze. This is indeed what happened; the team's next step will be to extract the plug of frozen lake water to examine it for microbial life.

The initial discovery of Lake Vostok prompted scientists to speculate about the life forms that might exist in this untouched world. John Priscu has no doubt that the lake, and indeed Antarctica's entire ice sheet, is alive. Having confirmed thriving microbial populations in the ice caps of smaller surface lakes, he also found viable bacteria in the Vostok ice core all the way down, lending support to the idea of a flourishing microscopic lake ecosystem that has not seen the light of day for more than 15 million years.

Apart from the Russian venture, there are now two other major drilling efforts under way. A team from the British Antarctic Survey is preparing to drill into the much smaller but almost equally deep Lake Ellsworth in western Antarctica, using no more than hot water to melt a bore hole for just long enough to sample the lake water and sediment.

In contrast, a US team supported by the National Science Foundation's Office of Polar Programs has chosen a very different subglacial ecosystem to study. Since the discovery of Lake Vostok, at least 200 subglacial lakes have been identified, and hundreds more are expected to be found. Most of the known lakes sit between 3 and 4 kilometres (1.7 to 2.5 miles) deep under the ice and are smaller than Vostok. Some are connected through an under-ice network of ancient, hidden rivers. Lake Whillans is connected with the ocean beneath the Ross Ice Shelf, and it drains and fills slowly but regularly. Its ebb and flow is thought to be one of the major influences on the movement of the ice shelf, and the US team plans to use hot-water drills to push through the 800 metres (0.5 miles) of ice above the lake and to deploy remote-controlled robots to explore the waterways below.

Whatever Antarctica's subglacial lakes will reveal, freshwater ecologist Ian Hawes is certain he has come as close as possible to an ancient world in Lake Untersee. This permanently ice-covered lake, in the Gruber Mountains of central Dronning Maud Land, is East Antarctica's largest surface lake. Diving under its 3-metre (10-foot, 3.2-yard) ice cap, Hawes and his US Antarctic Program colleagues discovered an eerie landscape of conical towers made by microbial mats.

The largest of the cones grow to half a metre (1.6 feet, 0.5 yards) and are laminated like plywood. Only the surface layer harbours actively metabolising bacteria, which grow on top of their predecessors like growth rings in a tree. Cyanobacteria are again the dominant group, but the lake is the only known place on Earth today where they construct such large and elaborate structures. However, three billion years ago, such mounds were one of the most common growth forms on the planet.

Hawes' team included astrobiologists from SETI, an institute dedicated to the search for extraterrestrial intelligence, who see Antarctica as an ideal place to study early life forms in the hope of gaining a better idea of what to look for on Earth-like planets elsewhere in the universe.

Coda **Beyond the Ice**

Antarctica is Earth's most other-worldly continent and the closest analogue for conditions on planets and moons in our own solar system or around more distant stars. It is also a perfect place from which to look far beyond our galaxy to the outer reaches and early days of the universe.

Above the South Pole, the sun sets only once a year. The polar night lasts six months, and thanks to the vortex of cold air that circles around this southernmost point of the planet, the atmosphere is more stable and transparent than anywhere else. From an astronomer's perspective, there is no better place on Earth to build an observatory to study some of the earliest and most energetic events in the universe and to track particles that are impossible to detect elsewhere.

Neutrinos have long been considered undetectable anywhere. They began life as a purely theoretical construct, a concept developed to fill a gap in existing laws and models that govern particle physics. When the young Austrian physicist Wolfgang Pauli proposed the existence of neutrinos in 1930 he did so almost apologetically, calling it a desperate solution and not expecting that anyone would be able to detect or measure the new particle. He had come up with the idea of a neutrino as a way of explaining a problem in nuclear physics, specifically in radioactive decay. It was known by that stage that the core of an atom could only exist in distinct energy levels and so, in theory, the radiation produced by radioactive decay should also be emitted in particular wavelengths, rather than the continuous spectrum that is actually measured. To make theory match reality, Pauli suggested that radioactive decay had another, as yet unseen, component in the form of streams of strongly penetrating, inconceivably small subatomic particles with a tiny mass and no charge. Pauli didn't like publishing papers, but instead disseminated his work by corresponding with other scientists. In the case of the neutrino, he announced his idea in a letter to the attendants of a conference on radioactivity, addressed 'Dear radioactive ladies and gentlemen'.

Each second, trillions of neutrinos pass though Earth, undetected and undeterred. Apart from photons, neutrinos are the most abundant particles in the

Begun in the early 1990s, AMANDA – or the Antarctic Muon and Neutrino Detector Array – preceded the IceCube project. Here scientists get ready to lower a neutrino detector into a hole in the ice. JOSH LANDIS/NSF

universe, but they have also proved the most elusive. It took a quarter of a century for them to be detected. Two years before Pauli died he received a telegram from an American physicist, Frederick Reines, with the news that the ghostly particles had finally been seen.

With no charge and only the tiniest mass, neutrinos pass through everything without interaction or scattering. Human bodies are no deterrent, and neither are planets. Of the billions of neutrinos that fly by every moment, the vast majority continue on their trajectory without disruption, and they remain undetectable. The only thing that can throw a neutrino off course is a direct collision with the core of an atom, but the chance of that happening is minute, and hence the challenge of capturing a neutrino in action is enormous. A neutrino hunter needs either an awfully large number of neutrinos or an awfully large detector. Reines and his team took the first approach.

Natural sources of neutrinos include the nuclear reactions in the centre of the sun, but they are also emitted by nuclear reactors. Reines had spent much of his career at Los Alamos, as a theoretical physicist on the Manhattan Project. Together with chemical engineer Clyde Cowan, he launched Project Poltergeist in 1951, hoping to use a bomb test as a source of neutrinos and two large vats of water to detect them. However, the plans changed and the team set up two hip-high water tanks outside the Savannah River reactor in South Carolina – and waited. The reactor produced 50,000 billion neutrinos per square centimetre (0.2 square inches) per second, but despite such enormous numbers, it would take about 20 minutes for any collisions to occur. The trick was to recognise when they did. Reines' plan was to detect gamma rays, the end product of a collision between a neutrino and a proton. It worked and eventually earned him and Cowan the Nobel Prize in Physics in 1995.

Once the capture of neutrinos seemed more feasible, other teams built even larger detectors deep inside old gold mines and caverns. A Russian-German team turned Lake Baikal in Siberia into a neutrino detector by lowering cables with light sensors deep into its clear water. In 1987, neutrino detectors in caves became the first to spot a few of the particles that sprayed out of a supernova called 1987A. However, it was soon clear that if the goal was to detect solar neutrinos in appreciable numbers, the instruments would have to be even more massive and shielded to avoid confusion with cosmic rays. Antarctica shifted into focus as an ideal site to ensnare neutrinos in the ice, and the construction of the first detector near the US Amundsen–Scott South Pole Station during the early 1990s paved the way for neutrino astronomy.

AMANDA (Antarctic Muon and Neutrino Detector Array) has since been superseded by IceCube, an even larger and more sophisticated upside-down telescope based on the same principle. When IceCube was completed in December 2010, it had transformed a cubic kilometre (0.3 cubic miles) of pristine polar ice into an enormous cosmic observatory. Construction teams had spent seven summers drilling 86 holes into the ice, each of them 2.5 kilometres (1.5 miles) deep, with a hot-water drill. Into each of these holes, they sank a string beaded with 60 round instruments the size of watermelons. In total, more than 5000 such exquisitely sensitive light detectors are now frozen in a three-dimensional grid in the ice, ready to catch neutrinos.

The pressure is so great at these depths that the ice is almost perfectly translucent. Whenever a neutrino smashes into an atom, the collision produces another

particle called a muon, which leaves a trail of blue light in its wake – and it is this flash of light that IceCube's sensors measure. If the light travels up, it is clear that it was produced by a neutrino because nothing else can travel through Earth unhindered.

The sun is an obvious first target for astronomers, as the neutrinos that are produced in its centre would allow them to probe what goes on deep inside the fireball. But the international team of physicists that uses IceCube has its sights set on much more distant and violent targets, including supernova explosions, black holes and cosmic rays. Because neutrinos make a beeline directly through the universe, they are thought to carry astronomical information from the most far-flung corners of the cosmos and from deep inside the most high-energy processes.

Cosmic rays were first observed by Victor Hess, an Austrian scientist, when he measured radiation from hot-air balloons at altitudes up to 5 kilometres (3 miles). These high-energy rays are now known to be mostly charged particles such as

Neutrino detectors, called digital optical modules (DOMs), sit on a shelf in the IceCube drilling building, awaiting deployment into the ice. PETER REJCEK/NSF

protons that originate in outer space and bombard our planet from all directions, at a rate of one per square metre (10 square feet) per day. Astronomers think that some of these protons are accelerated by supernova explosions, but the source of the strongest cosmic rays, thousands of times more energetic than anything that can be produced on Earth, remains one of the biggest mysteries in physics.

So far it has been impossible to track cosmic rays back to their origin in space because charged particles are deflected by magnetic fields that are woven through the universe near galaxies. Not so neutrinos. The same processes that produce cosmic rays also release high-energy neutrinos, and IceCube should help astronomers to locate regions where they have been generated. Over time, the observatory will build up a neutrino map of the sky and identify hotspots where high-energy cosmic rays appear to be coming from.

Nearby, at the American polar base, the 10-metre (30.8-foot) South Pole Telescope is tracking the evolution of the early universe, with the aim of unravelling a

mysterious ingredient known as dark energy. Astronomers agree that the visible part of each galaxy or nebula is only the smallest component and that all known matter accounts for no more than 4 per cent of the universe. The rest is divided into two other major cosmic components, dark matter and dark energy, which remain invisible to any method of observation that has so far been developed.

Of the two, dark energy is the more mysterious, and one of the main goals since the launch of the telescope in 2007 has been to confirm its existence. Dark energy is considered an opposite force to gravity, pulling things further apart rather than pushing them closer together. If it exists, it should change how clusters of galaxies, some of the biggest things in the universe, are thought to grow over time.

As a first step, the telescope has begun mapping the skies for such clusters. Unlike optical or radio telescopes, it probes the universe with wavelengths that

are longer than infrared but shorter than radio waves, making use of the fact that clusters affect the cosmic microwave background, the remnant radiation that has been travelling through space since the Big Bang. Anywhere else in the world, such measurements would be impossible because the atmosphere holds too much water vapour and clouds the view of this ancient trace left behind by the birth of the universe.

Antarctica is not just the best but the only location for either of these ambitious astronomical ventures, and the scientists involved expect that they will open a window on the farthest reaches of the universe and, in the process, discover places we didn't even know existed. For now, the stage is set for a new Heroic Era of exploration upon the frozen continent – into both the infinite vastness of space and the fabulous microcosm subsisting in and beneath the ice.

An aurora above the South Pole Telescope, which collects data on cosmic microwave background radiation and dark matter. The red light indicates that someone is working on the telescope. KEITH VANDERLINDE/NSF

Sources and Further Reading

This book is partly based on interviews and less formal conversations, some of which began during my first visit to Scott Base a decade ago and have been picked up again over the years as part of my coverage of Antarctic research. However, several books, as well as websites and databases, have been invaluable during the process of writing.

Two books in particular have accompanied me throughout this project. David Harrowfield's *Call of the Ice* (Auckland, David Bateman, 2007) was published to mark the 50th anniversary of Scott Base and New Zealand's Antarctic science programme. It could only have been written by someone with a long and close association with Antarctica, and it provides the most reliable and comprehensive descriptions of life and work on the ice. Apart from detailed summaries of research projects, Harrowfield also examines education initiatives and the impact of Antarctic tourism.

In many ways, *Call of the Ice* picks up where *Antarctica* (New York, Praeger, 1965), edited by Trevor Hatherton, leaves off. Each chapter in this book is written by a scientist at the forefront of a particular discipline and, collectively, the entries give a thorough overview of New Zealand's science effort during the International Geophysical Year. Yorkshire-born Hatherton was a geophysicist, and it was his interest in Antarctica that led him to apply for the first National Research Scholarship to be offered to an overseas scientist wanting to work in New Zealand. He arrived in 1950. Five years later he travelled to Antarctica for a reconnaissance on foot across the sea ice of McMurdo Sound to select a site for New Zealand's base. From 1956 to 1958, he led New Zealand's science team during the International Geophysical Year. Hatherton was passionate about science, and likened the research process to 'looking for a needle in the haystack and finding the farmer's daughter'.

Introduction A Land of Ice

Pages 1–5: For many years, Russian scientists were poised to become the first to enter one of Antarctica's subglacial lakes. The announcement that they had indeed reached the bottom of the ice sheet and the surface of Lake Vostok came on February 8, 2012, issued by the Arctic and Antarctic Research Institute (AARI): www.aari.nw.ru/docs/press_release/2012/Lake%20Vostok%20080212.html

The achievement was also covered by several scientific journals, including *Nature* and *Science*.

Pages 9–11: Creina Bond and Roy Siegfried deliver a very readable account of Antarctic exploration and the exploits of the sealing and whaling industries in *Antarctica: No Single Country, No Single Sea* (London, Hamlyn, 1979) – including some unexpected connections: 'Adolf Hitler changed the fertility of Antarctic whales. As his regiments began their march across Europe, whaling ships withdrew from the Southern Ocean . . . . It was the first time in more than 30 years that the whales had been left to themselves.'

In relation to Antarctica, James Cook's second voyage (1772–75) was the most significant. He commanded HMS *Resolution*, with Tobias Furneaux in charge of the expedition's companion ship HMS *Adventure*, and crossed the Antarctic Circle three times. New Zealand historian John Cawte (known as J. C.) Beaglehole edited Cook's journals from all three voyages. The project was funded by the Hakluyt Society, and between 1955 and 1967 the massive volumes of *The Journals of Captain Cook on his Voyages of Discovery* (Cambridge University Press) appeared. Volume II, published in 1961, focuses on Cook's exploration of the Southern Ocean, during which he sailed as far as 71 degrees South.

While Cook's own journals offer the most direct insight, another major work resulted from his second expedition. Johann Reinhold Forster, who together with his son Georg was taken on as the scientist for the voyage, also produced his own account, amid controversy. *Observations made during a Voyage round the World, on Physical Geography, Natural History, and Ethic Philosophy* (London, G. Robinson, 1778) focuses on the scientific discoveries.

James Clark Ross's dedication to marine science comes through on almost every page of his expedition diary, *A Voyage of Discovery and Research in the Southern and Antarctic Regions* (two volumes, first published in London by John Murray in 1847, but available through David & Charles Reprints, 1969), which is also worth browsing for Joseph Dalton Hooker's drawings.

Every expedition that took place during the subsequent Heroic Era of Antarctic exploration produced several publications of journals, paintings and sometimes photographs. The Antarctic Heritage Trust (New Zealand) keeps a comprehensive list of primary and secondary sources (www.nzaht.org/AHT/FurtherReading).

For this book, I have focused on the expeditions led by Scott and Shackleton. *The Voyage of the 'Discovery'* (two volumes, London, Smith, Elder & Co., 1905) is Scott's journal from his first Antarctic expedition, officially known as the British National Antarctic Expedition 1901–04. Apart from the extensive descriptions of the scientific discoveries, this journal also introduces many men who would become leading figures in Antarctic exploration, including Shackleton.

Scott was arguably one of the most prolific and thoughtful diarists of the Heroic Era. In Volume I of *Scott's Last Expedition* (London, MacMillan & Co., 1913), the journal of the Terra Nova expedition, the tragedy of the polar party lends his letters to friends and relatives of those who died with him an intensity that remains unmatched by any other expedition account. The second volume was written by members of the expedition. It includes Apsley Cherry-Garrard's diary of the winter journey to Cape Crozier as well as Victor Campbell's account of the northern party, which survived an Antarctic winter stranded in an ice cave.

1 Uncovering the Past

Epigraph text by Robert McCormick, surgeon on HMS *Erebus*, from James Clark Ross's diary, *A Voyage of Discovery and Research in the Southern and Antarctic Regions*, Appendix IV (geological remarks on the Antarctic continent and southern islands), p. 413.

Pages 22–23: *The Crossing of Antarctica* (London, Cassell, 1958) by Sir Vivian Fuchs and Sir Edmund Hillary is the story of the Commonwealth Trans-Antarctic Expedition, written by its leaders. Sir Edmund's chapters recount his Ferguson tractor journey to Cape Crozier, undertaken as a test run to determine whether the vehicles would perform well enough to be useful on the trip south.

Hellbent for the Pole (Auckland, Random House, 2007) by Geoffrey Lee Martin, the journalist who accompanied the expedition to cover it for the New Zealand Press Association, the *New Zealand Herald* and the *Daily Telegraph* in London, was written half a century after the crossing. It zooms in on the furore Hillary and four of his companions caused when they drove their modified Ferguson farm tractors to the South Pole and beat Fuchs and his more sophisticated Sno-Cats.

Pages 41–47: *Antarctic Climate Change and the Environment*, a special report published by the Scientific Committee on Antarctic Research (SCAR) as a contribution to the International Polar Year 2007/08 (free to download at: www.scar. org/publications) gives a comprehensive update on how the physical and biological environment of the Antarctic continent and the Southern Ocean has changed from deep time to the present.

Pages 24–53: Peter Barrett and Lionel Carter both contributed chapters to *Antarctic Climate Evolution*, the eighth volume in a series of books called *Developments in Earth and Environmental Sciences* (edited by Fabio Florindo and Martin Siegert, Amsterdam, Elsevier, 2008).

Pages 53–54: The Belgica expedition (1897–99) was a modest venture focused on science. Inadvertently, its members became the first people to endure an Antarctic winter. Among them were Frederick Albert Cook, who would later become embroiled in a controversy over the discovery of the North Pole, and Roald Amundsen, who, frustrated in his wish to reach the North Pole, redirected his focus to the South Pole. Cook's *Through the first Antarctic Night 1898–1899* (New York, Doubleday & McClure, 1900) is a narrative of the *Belgica*'s voyage.

Pages 55–56, 59: The secondary literature about Scott and Shackleton and their respective leadership qualities is vast. One of the best-known authors is Roland Huntford, who wrote biographies of Scott, Shackleton, and Norwegian polar explorer Fridtjof Nansen. In his controversial *The Last Place on Earth: Scott and Amundsen's Race to the South Pole* (New York, Modern Library paperback edition, 1999) Huntford argued that Scott made several errors in judgement (including the reliance on man-hauling instead of sled dogs) that ultimately led to his failure.

In her book *The Coldest March* (New Haven, Yale University Press, 2001) Susan Solomon brought a scientific perspective to the debate about Scott's legacy. By

OPPOSITE The crew of HMNZS *Endeavour* throw blocks of snow on board to replenish the water supply during the 1957/58 Trans-Antarctic Expedition. ANTARCTICA NZ PICTORIAL COLLECTION: UNKNOWN/TAE 461 analysing extensive meteorological records, she came to the conclusion that Scott's polar party suffered exceptionally cold conditions, which thwarted their effort despite their meticulous predictions and preparations.

2 Life on Ice

Apsley Cherry-Garrard was Scott's major literary rival. It was he who urged the Tennyson inscription from 'Ulysses' for the polar party's memorial cross. *The Worst Journey in the World* (originally published in London, by Constable in 1922, but later paperback editions in New York, by Carroll & Graf, 1989, 1997, 2001) has all the elements of high drama, but remains eminently readable. Unlike many other official accounts and journals, which were rushed into print to pay off debt after the expedition, Cherry-Garrard's book is the product of considerable research and thoughtful writing.

He was painfully shy and prone to bouts of depression, yet also determined and passionate. Sara Wheeler's biography *Cherry: A Life of Apsley Cherry-Garrard* (London, Jonathan Cape, 2001) illuminates his character before and after his epoch-making polar expedition. Wheeler's earlier travelogue *Terra Incognita: Travels in Antarctica* (London, Jonathan Cape, 1996) entertainingly communicates the icy spell Antarctica holds over her and most other people who have a chance to travel south.

Pages 70–83: Several academic books have been published on the biology and evolution of Antarctic fish. One of the more readable ones, *Antarctic Fish Biology: Evolution in a Unique Environment* (New York, Academic Press, 1993), was written by Joseph T. Eastman, a pioneer in the field.

For the more taxonomically minded, *Fishes of the Southern Ocean* (edited by O. Gon and P. C. Heemstra, published by the J. L. B. Smith Institute of Ichthyology, Grahamstown, South Africa, 1990) is a comprehensive account of fish taxonomy, but it also includes chapters on the origin, evolution, systematics, biology, conservation and exploitation of Antarctic fishes.

As mentioned above, *A Voyage of Discovery and Research in the Southern and Antarctic Regions*, James Clark Ross's diary of his nineteenth-century expedition, is an enthralling read.

Pages 84–96: Antarctica's underwater world is the topic of the last chapter of *The Living Reef: The Ecology of New Zealand's Rocky Reefs* (edited by Neil Andrew and Malcolm Francis, Nelson, Craig Potton Publishing, 2003). This chapter is only a few pages long, but it is a great introduction to the invertebrate life forms that thrive under the sea ice.

Pages 104–118: *Skua and Penguin: Predator and Prey* (Cambridge University Press, 1994) is Euan Young's own account of his research. Although it is an academic book, its main characters and Young's fresh writing style make it very readable and enjoyable.

Meredith Hooper is a writer with several Antarctic titles to her name. In *The Ferocious Summer: Palmer's Penguins and the Warming of Antarctica* (London, Profile Books, 2007), she charts her summer at Palmer Station, where she joined scientists trying to understand the causes and impact of a fast-changing climate on resident penguin colonies. Her most recent book, *The Longest Winter: Scott's other Heroes* (London, John Murray, 2010), recounts the great adventure of Scott's northern party. Scott's ship, the *Terra Nova,* dropped off six men in Terra Nova Bay for what had been planned as a six-week exploration of that stretch of Victoria Land. However, ice conditions were such that the ship never made it back, and the men were forced to live through the polar night in a snow cave, subsisting on seal meat. They survived, and man-hauled their sledge across more than 200 kilometres to return to base at Cape Evans, but their heroic tale is often overshadowed by the tragedy of the polar party.

3 True Antarcticans

Alastair Fothergill's *Life in the Freezer: A Natural History of the Antarctic* (London, BBC Books, 1993) is an excellent introduction to how animals and plants survive in Antarctica. It was written to complement a television series of the same name. Epigraph text used courtesy of Sir David Attenborough.

Pages 124–131: *Life at the Limits: Organisms in Extreme Environments* (Cambridge University Press, 2002) is a guided tour through a number of habitats that many of us would expect to be lifeless. David Wharton explores various survival strategies that animals and plants have evolved to deal with extreme temperatures at either end of the scale, desiccation, high salinity, pressure, long periods of darkness or extreme intensity of light – and he asks what life may look like elsewhere.

For anybody interested in nematodes in particular, David Wharton and R. N. Perry edited *Molecular and Physiological Basis of Nematode Survival* (Wallingford, CABI Publishing, 2011), which also includes a chapter on cold tolerance, written by Wharton.

Pages 134–141: 'Antarctica is the coldest, driest, highest and windiest continent, its plants grow where it is warm, wet, low and calm. Ecophysiologists should be grateful.' This introduction to Allan Green's chapter on plant life in Antarctica, in *Functional Plant Ecology* (Almeria, CSIC Press, second edition, 2010), is a good indication of the readability of the rest of the chapter, possibly the most enjoyable in this textbook.

The Illustrated Moss Flora of Antarctica (edited by Ryszard Ochyra, Ronald I. Lewis Smith and Halina Bednarek-Ochyra, Cambridge University Press, 2008) is a comprehensive identification guide, but in its first 50 pages this book also covers the history of botanical expeditions, the evolutionary origins of Antarctica's moss flora, and the ecology of how mosses survive on the frozen continent.

Pages 140–146: During the Heroic Era of Antarctic exploration (1895–1915), four expedition parties built bases in the Ross Sea region of Antarctica. They still stand today, and in 2002 the Antarctic Heritage Trust embarked on a long-term conservation project to protect their legacy. Conservators stay on the ice throughout the year, and they blog about their project on the Natural History Museum website (www.nhm.ac.uk/natureplus/community/antarctic-conservation).

During the peak of the short Antarctic summer, ice thaws on the surface of a lake in the Garwood Valley.
ROB McPHAIL

Pages 147–154: In *Full House: The Spread of Excellence from Plato to Darwin* (New York, Harmony Books, 1996) Stephen Jay Gould challenges the notion that humankind is the crowning achievement of evolution. Instead he argues that microbes are, and always have been, the dominant forms of life on Earth.

Microbiology of Extreme Soils (Berlin, Springer Verlag, 2008) features a brief chapter on Antarctica's hot zones, while *Antarctic Microbiology* (New York, John Wiley & Sons, 1993) explores marine, terrestrial and freshwater environments and the microbes that survive in each.

4 Oasis in a Frozen Desert

The Voyage of the 'Discovery', Scott's journal from his first Antarctic expedition, covers the discovery of the Dry Valleys from page 292.

Pages 159–174: Colin Bull wrote *Innocents in the Dry Valleys: An Account of the Victoria University of Wellington Antarctic Expedition 1958–59* (Wellington, Victoria University Press, 2009) half a century after the event, but it reads as if he were still a young man, ready to embark on a big adventure. Written in the same humorous style he used in speeches and interviews, this book recounts the beginnings of what has since turned into a comprehensive Antarctic research programme at Victoria University. It also includes significant contributions by his companions Peter Webb, Richard 'Dick' Barwick and Barrie McKelvey, whose names can all be found on landscape features in Antarctica's arid desert. Bull later became director of the Institute of Polar Studies at Ohio State University in 1965, during which period he sent the first female scientists to Antarctica. He has also written *Innocents in the Arctic: The 1951 Spitsbergen Expedition* (Fairbanks, University of Alaska Press, 2005) about his first polar expedition.

Pages 176–201: Bill Green's *Water, Ice and Stone: Science and Memory on the Antarctic Lakes* (New York, Bellevue Literary Press, 2008) is one of the most literary books on Antarctic research. Green succeeds in conveying what motivates him as a scientist exploring the permanently ice-covered lakes in the Dry Valleys on both a personal and professional level. The book is an account of his experiences during a field season, interwoven with personal thoughts and meditations on the stunning landscape. His prose has been described as 'geochemistry made into geopoetry', and this is certainly a book that can be enjoyed for the love of words.

Green also wrote an essay for *Improbable Eden: The Dry Valleys of Antarctica* (Nelson, Craig Potton Publishing, 2003) to accompany Craig Potton's stark images of one of Antarctica's most extraordinary landscapes.

Scientific publications

Most people who travel to Antarctica are scientists, and their professional currency is publication in scientific journals. Rather than listing all the papers I have consulted, here are links to some of the umbrella organisations that support Antarctic research, most of which have easily accessible databases of published research:

National Science Foundation Office of Polar Programs – www.nsf.gov/div/index.jsp?div=ANT
British Antarctic Survey – www.antarctica.ac.uk
Scott Polar Research Institute – www.spri.cam.ac.uk
Scientific Committee on Antarctic Research (SCAR) – www.scar.org
Australian Antarctic Division – www.antarctica.gov.au
Antarctica New Zealand – www.antarcticanz.govt.nz
National Institute of Water and Atmospheric Research – www.niwa.co.nz
GNS Science – www.gns.cri.nz

Supplementary sources

In addition to the publications listed above, sources of more general influence on the writing of this book include the following works of photography and poetry.

Antarctic photography
Frank Hurley was an Australian photographer who first travelled to Antarctica at the age of 23, as part of Douglas Mawson's expedition. However, he is best known for the images he produced during Shackleton's 1914 expedition, which was marooned for almost two years. Hurley compiled his records into the 1919 documentary film *South* and his footage was also used in the 2001 film *Shackleton's Antarctic Adventure*.

Herbert Ponting was Scott's expedition photographer and cinematographer. *With Scott to the Pole: The Terra Nova Expedition 1910–1913* (London, Allen & Unwin, 2004) brings together some of his best images from the archives of the Royal Geographical Society and the Scott Polar Institute.

New Zealand mountaineer and photographer Colin Monteath spent 29 summers on ice, co-ordinating field operations and logistic support. Following the 1979 Air New Zealand crash on the slopes of Mount Erebus, he helped with the recovery operation. *Antarctica: Beyond the Southern Ocean* (Auckland, David Bateman, 1996), *Vanishing Wilderness of Antarctica* (Novara, White Star, 2010) and *Antarctica: Land of Silence* (Auckland, David Ling Publishing, 2010) bring together some of his images from the frozen continent.

Craig Potton's *Improbable Eden* and *White Silence: Grahame Sydney's Antarctica* (Auckland, Penguin, 2008) provide plenty of inspiration, as does *Still Life: Inside the Antarctic Huts of Scott and Shackleton* (Sydney, Murdoch Books, 2010), a photographic tribute to Antarctic heritage by Jane Ussher.

Antarctic poetry
It would go beyond the scope of this book to list all volumes of poetry that focus on Antarctica. The following is a select list of New Zealand works that have been a source of inspiration: Bill Manhire's *Collected Poems* (Wellington, Victoria University Press, 2001), which includes a section of Antarctic Field Notes; Bernadette Hall's *The Ponies* (Wellington, Victoria University Press, 2007); Bill Sewell's *Erebus: A Poem* (Christchurch, Hazard Press, 1999); and Chris Orsman's *South* (Wellington, Victoria University Press, 1996) and *Black South* (Wellington, Pemmican Press, 1997).

Orsman, Manhire and painter Nigel Brown also produced *Homelight: An Antarctic Miscellany* (Wellington, Pemmican Press, 1998) during their visit as the inaugural New Zealand Antarctic arts fellows. Only a handful of editions were produced at Scott Base, the selection borrowing its title from the brand of lamp oil Scott took on his fatal expedition to the South Pole.

The Wide White Page (edited by Bill Manhire, published in Wellington, Victoria University Press, 2004) is a collection of imaginative, fictional accounts of Antarctica, mostly by authors who have travelled south only in their mind.

Acknowledgements

My first thanks go to the people who got me to the ice. Journalists have accompanied scientists to Scott Base since the International Geophysical Year in 1957, and today Antarctica New Zealand's media programme still provides an extraordinary opportunity for those interested in science and life at the world's most isolated research station. I felt privileged when I was invited to visit Scott Base and some of the remote field camps in 2001, thinking that this would be a once-in-a-lifetime journey. But Antarctica got under my skin and I wanted to return. Five years later I did, again with the help of Antarctica New Zealand, and I hope it was not my last visit. Everybody at Antarctica New Zealand has been extremely helpful throughout this book project, but in particular I would like to thank Lou Sanson, Ed Butler, Shulamit Gordon, Matt Vance, Paul 'Woody' Woodgate, and Tom Riley, as well as former staff Ursula Ryan, Natalie Cadenhead, and Dean Peterson.

After more than fifteen years of covering science, I have yet to meet a scientist who actually fits the white-coated stereotype. In my experience, scientists are passionate about their work, generous with their time and knowledge, and keen to make sure others understand what they do and why they do it. The Antarctic science community is a special part of the larger group. They have an additional attraction to offer – the ultimate field trip. For everybody mentioned by name in this book, there are others who supported me by explaining their research, sending me manuscripts of yet-to-be published papers and book chapters, putting me in contact with other research groups, or simply plying me with stories and the occasional drink with ancient ice cubes. This book builds on my coverage of Antarctic science over the years, and it is no exaggeration to say that hundreds of scientists have contributed to it, in one way or another. I am grateful – and if there is anybody who feels I have overlooked their work, I offer my apologies. Without the generosity, patience and support of the science community, I would have no stories to tell.

Similarly, without Radio New Zealand, I would have no audiences for my stories. I have produced and presented the broadcaster's science programme for more than a decade and haven't had a moment of boredom. My managers have supported me as the show reinvented itself several times during this period, and I would like to thank Paul Cavanagh, Paul Bushnell, Phil Smith, and all my colleagues in the features department.

Secrets of the Ice would not be the book it is without the photographers who contributed to it. Most of them have a day job and pursue photography as a passion.

Rob McPhail can fly and land a helicopter almost anywhere. He has spent more than twenty summers flying science teams in and out of remote camps and opened up the vast library of extraordinary images he collected en route. Peter Marriott and Rod Budd are professional divers and scientists and among the few people who get to see – and photograph – Antarctica's colourful underwater world. Antarctica New Zealand provided many images from its collection, which in turn was assembled by the scientists themselves; and Peter West, at the US National Science Foundation's Office of Polar Programs, introduced me to their image archives. The Antarctic Heritage Trust, GNS Science, NIWA, Geometria, and many individual scientists have freely provided images and graphics to illustrate this book. Lionel Carter (Victoria University) and Simon Cox (GNS Science) provided invaluable help with the design of graphics and maps.

I could not have asked for a better originating publisher than Auckland University Press. Sam Elworthy was extremely patient throughout the process. As I kept asking for yet another extension of the deadline, he remained encouraging and positive. His feedback has been invaluable – always constructive, insightful and knowledgeable – and interspersed with reassuring advice about how to survive a first year of parenting. At Yale University Press, Joseph Calamia was always enthusiastic and supportive. Mike Wagg takes the credit for much that is good about the manuscript. His copy-editing was both fast and accurate – a rare combination indeed. Anna Hodge and Katrina Duncan were a great team to work with. They took the raw ingredients, applied their magic and created the book you now hold in your hands. My thanks also go to Louise Belcher for proofreading, Ginny Sullivan for indexing and Spencer Levine for designing the maps.

Further support and inspiration has come from many corners. Mountaineer and photographer Colin Monteath probably doesn't realise how much of a quiet inspiration he (and his incredible polar library) has been over the years. Harry Ricketts, at Victoria University's International Institute of Modern Letters, and my fellow creative writing students suffered through several early drafts – unharmed, I hope. The idea for this book was born a year or so before our son Lukas and, time and again, life buoys appeared in the maelstrom that is sleep-deprived, first-time parenting through the generosity and support of family and friends. The new grandparents flew halfway across the globe to help with childcare; and friends fed and watered us when there was no energy left for anything, and looked after Lukas when I thought I would write, but actually needed to sleep.

My main support team, however, is my husband, Andy, and Lukas. For months, Andy took on more than his fair share of childcare and home duties, while I was hiding in the office and of little use to anybody. He never complained. On the contrary, he encouraged me to keep going, even at times when I wasn't so sure about it myself. And Lukas sat on my lap for hours, 'helping', as I was sifting through hundreds of images. As a result, aged two, he can already tell the difference between emperor and Adélie penguins. Without Andy and Lukas, my explorations of science on ice would not have been possible. This book is dedicated to them.

Index

Numbers in **bold** indicate images and captions.

Cape Crozier, 15, 63, 109, 112–13, 119, 211, 213

Cape Evans, **11**, 15, 85, 88, **90**, 91, 92, 107, 141, **144**, 165, 215

Cape Hallett, **132**, 133, **139**

Cape Roberts drilling project, 10, 30, 31, 32–33, 40, 44

Cape Royds, **11**, 15, 100–1, 101, **102–3**, **106**, 107, 109, 112, 116, 117, 119

carbon, 29, 53, 91, 93, 136, 169–71, 176, 181, 188, 189, 195

carbon dioxide, 32, 33, 39, 40, 46, 47, 95, **173**, 174, 175

Carter, Lionel, 47, **51**, 213

Cary, Craig, **148**, **151**, **152–3**, 166, 167–70, 167, 168, 170, 173

Census of Antarctic Marine Life (CAML), 97

Cheng, Chris, 76–77

Cherry-Garrard, Apsley, 63, 211, 214

China, 26, 143; and scientific study of Antarctica, 46

chirpsounder, *see* ionosonde

chlorine, 54–56, 142

chlorine dioxide, 55–56

chlorofluorocarbons (CFCs), 54, 56, 59

chlorophyll, **196**

chloroplast, 174

climate change, 19, 22, 32–33, 40–43, 47, 53, 59, 83, 95, 119, 213

climate, research on and shifts in, 1, 4, 6, 11, 19, **20–21**, 22, 27, 29–30, 33, 37, 40–47, 48, 52–53, 192, 199, 215

clothing, at Antarctica, 41, **42**, 54, **151**, 169

Cloudmaker, **155**

clouds, polar stratospheric, 54, 55–56

coal and coal beds, 25, 27

Colorado State University, 124

Columbia University, 47

Commonwealth Trans-Antarctic Expedition, 22, 23, 28, 213

continental drift, 26, 27

Cook, Captain James, 9, 211

corals, 88

Coulman High, 40

Cowan, Clyde, 205

Cowan, Don, 168, 168

Cozzetto, Karen, **194**

crabs, 88, 96

Crary Lab, McMurdo Station, 37

crustaceans, 83, 184

Crutzen, Paul, 56

cryoconites, 183, 193

cryptobiosis, 124

Cummings, Vonda, 92–93, 92, 95

currents, Antarctic, 6, 19, 22, 27, 47, 48, 49, 50, 50, **51**, 72, 88, 91, 98, 116, 187; *see also* Antarctic Circumpolar Current

cyanobacteria, 174–6, 181, **182**, 184, 187–9, 190, 192–3, 197, 199, 201

Daniel, Roy, 146–47, **149–50**, 154

Dansgaard–Oeschger (D-O) events, 47

Darwin Glacier, 184

Dayton, Paul, 88

Deep Sea Drilling Project, 28–29, 32

deglaciation, 51

Department of Scientific and Industrial Research (DSIR), 175, 182, 184, 185, 192

desert, 6, 11, 25, 32, 63, 124, 125, 143, 161, 163, 167, 171–2, 175, 187, 216

DeVries, Art, 73, 74, 75–76, 78, 79, 81, 82, 86

diatomite, 39

dinosaurs, 19, 26, 27

Discovery expedition (1901–04), **9**, 14, 23, 101, 159

Discovery, HMS, 86

Dissostichus mawsoni, **98**; *see also,* toothfish, Antarctic

diving: in Antarctic lakes, 63, 185–88, 186–87, 201; under sea ice, 83–86, 84, 85, 86, 87, 88, **89**, **94**, 95, 185

DNA analysis and fingerprinting, 71, 114–15, **148**, 150, 151, 166, 197; *see also* genetics

Dobson spectrophotometer, 56

Dobson, Gordon, 56

Dome A, 14, 46

Dome C, 14, 41, 46

Dome Fuji, 14, 42

Don Juan Pond, **195**

Drake Passage, 71

drilling: ocean, 28–31, **31**, 34, **34**, **35**, 36, **36**, **38**; sedimentary, 31, 33, 37, 41–47, **42**, **43**; subglacial lakes, 1, 4–5, 201; *see also* Antarctic Geological Drilling project (ANDRILL); Cape Roberts drilling project; Deep Sea Drilling Project

Dronning Maud Land, 14, 201

Dry Valleys, *see* McMurdo Dry Valleys

Drygalski Ice Tongue, 116

Drygalski, Erich von, 86

Dumont d'Urville, Jules Sebastien, 105

E. coli, 154; *see also* bacteria

East Antarctic Ice Sheet, 6, 14, 30, 33, 37, **155**

East Antarctic Plateau, 42

echinoderms, 94

ecologists, 73, 83, 88, 92–93, 92, 100, 117, 118, **155**; freshwater, 175, 201

ecosystems: freshwater, 184, 192, 201; marine, 92–95; protection of, 186, 201

eelpouts, 82

elegant sunburst lichen, *see Xanthoria elegans*

Ellsworth Mountains, 14, 24

Endeavour, HMNZS, 28, 212

endosymbiosis, 174; *see also* symbiosis

ENDURANCE robot, 195, **196**

Erebus Bay, **52**

Erebus, HMS, 9, 70, 213

European Space Agency, 129

Evans, Clive, 76–78

Evans, Edgar, 160

evolution, study of in Antarctica, 4, 5, 19, 70–73, 76, 79, 81, 82, 83, 89, **113**, 114–17, 151, 154, 174, 176, 182, 199, 207, 214, 215, 216; *see also* DNA analysis and fingerprinting; fossils and fossil record; palaeontologists

Falkland Plateau, 50

Farman, Joseph, 54–55

Farrell, Roberta, 141–7, 143

fauna, 124, 132, 154; invertebrate, 93, 123–4, 128, 132, **155**; vertebrate, 26, 71–72, 83, 123; *see also names of individual species*

Ferrar dolerites, **26**, **174**

Ferrar Glacier, 15, 19, 23, 140, 159, 160

Ferrar, Hartley Travers, **14**, **159**

fish, Antarctic, 11, 70–73, 75, 76–83, 89, 97, 214; *see also names of individual species*

flora, 33, 134–**8**, **215**; *see also names of individual species*

Florida State University, 37

foraminifera, 28–30

forams, *see* foraminifera

fossil fuels, use of, 24, 32

fossils and fossil record, 26, 28–29, 37, 40, 71, 114–15, 151, 197, 199

freeze-drying, 169

Fuchs, Vivian, 22–23, 213

fungi, 133, 136, 140–46, 154, 169–70, 173; *see also Cadophora*

Gaia theory of Earth, 54

Gardiner, Brian, 54–55

gastropods, 97–98, 184

Gauss (expedition ship), 86

genetics, 76, **113–17**, 119, 140, **148**, 151, 154, 166, 167, 170, 174, 192–3, 197; *see also* DNA analysis and fingerprinting

geochemists and geochemistry, 47, 176, 182–3

geologists and geology, 4, 11, 14, 22–26,

First published in the United States in 2012 by Yale University Press.
First published in New Zealand in 2012 by Auckland University Press.

Yale University Press books may be purchased in quantity for educational, business, or
promotional use. For information, please e-mail sales.press@yale.edu (U.S. office) or
sales@yaleup.co.uk (U.K. office).

Printed in China by 1010 Printing International Ltd.

Library of Congress Control Number: 2012935088
ISBN 978-0-300-18700-7 (hardcover)

A catalogue record for this book is available from the British Library.

Jacket, front: the frozen ocean meets the ice-covered mountains that form the western
boundary of Granite Harbour. Peter Marriott, NIWA. Pages ii–iii: a remote field camp on
the Darwin Glacier; jacket, back, and pages iv, 60: emperor penguins on sea ice; pages 16,
120: ice effects on the slopes of Mount Erebus; and page 156: a campsite in the Garwood
Valley. All by Rob McPhail. All uncredited photographs were taken by the author.

10 9 8 7 6 5 4 3 2 1